Methods for the Examination of Organismal Diversity in Soils and Sediments

CAB INTERNATIONAL is an intergovernmental organization providing services worldwide to agriculture, forestry, human health and the management of natural resources.

The information services maintain a computerized database containing over 2.7 million abstracts on agricultural and related research with 150,000 records added each year. This information is disseminated in 47 abstract journals, and also on CD-ROM and online. Other services include supporting development and training projects, and publishing a wide range of academic titles.

The four scientific institutes are centres of excellence for research and identification of organisms of agricultural and economic importance: they provide annual identifications of over 30,000 insect and microorganism specimens to scientists worldwide, and conduct international biological control projects.

International Mycological Institute
An Institute of CAB INTERNATIONAL

The Institute, founded in 1920, and employing over 70 staff:

- Is the largest mycological centre in the world, and is housed in a complex specially designed to support its requirements;
- Carries out research on a wide range of systematic and applied problems involving fungi (including lichens and yeasts), bacteria, and also on the preservation of fungi, the biochemical physiological and molecular characterization of strains, and on crop protection, environmental and industrial mycology, food spoilage, public health, biodeterioration and biodegradation;
- Provides an authoritative identification service, especially for microfungi of economic and environmental importance (other than certain human and animal pathogens), and for plant pathogenic bacteria and spoilage yeasts;
- Undertakes extensive computerized and indexing bibliographic work, including the preparation of 7 serial publications;
- Supports advice and undertakes a wide variety of project and culture work on crop protection, environmental, food spoilage and industrial topics.
- Offers training in pure and applied aspects of mycology and bacteriology, both at the Institute and overseas.

The Institute's dried reference collection numbers in excess of 355,000 specimens representing about 31,500 different species, and a genetic resource collection holds more than 17,000 living isolates by a variety of the most modern methods. The library receives about 600 current journals, and has extensive book and reprint holdings reflecting the Institute's interests.

International Mycological Institute
Bakeham Lane
Egham
Surrey TW20 9TY
UK

Tel: (01784) 470111
Telex: 9312102252
Fax: (01784) 470909
E-mail: CABI-IMI@CABI.ORG

For full details of the services provided, please contact the Institute Director.

Methods for the Examination of Organismal Diversity in Soils and Sediments

Edited by

Geoffrey S. Hall
International Mycological Institute
Egham, Surrey TW20 9TY, UK

Project coordinators

Pierre Lasserre (UNESCO)

and

David L. Hawksworth (IUBS)

CAB INTERNATIONAL

in association with
United Nations Educational, Scientific and Cultural Organization (UNESCO)
and the International Union of Biological Sciences (IUBS)

CAB INTERNATIONAL
Wallingford
Oxon OX10 8DE
UK

Tel: +44 (0)1491 832111
Fax: +44 (0)1491 833508
E-mail: cabi@cabi.org

CAB INTERNATIONAL
198 Madison Avenue
New York NY 10016-4341
USA

Tel: +212 726 6490
Fax: +212 686 7993

Published in association with:

International Union of Biological Sciences (IUBS)
51 boulevard de Montmorency
75016 Paris
France

and

United Nations Educational, Scientific and Cultural Organization (UNESCO)
7 place de Fontenoy
75352 Paris
France

A catalogue record for this book is available from the British Library, London, UK
A catalogue record for this book is available from the Library of Congress, USA

ISBN 0 85199 149 1

Also available in the IUBS Series of Methodology Handbooks and published by CAB INTERNATIONAL:

Tropical Soil Biology and Fertility: A Handbook of Methods, 2nd Edn
Edited by J.M. Anderson and J.S.I. Ingram
1993 240 pages ISBN 0 85198 821 0

Biological Monitoring of the Environment: A Manual of Methods
Edited by J. Salanki, D. Jeffrey and G.M. Hughes
1994 176 pages ISBN 0 85198 893 8

Printed and bound in the UK at the University Press, Cambridge

Contents

Foreword

Our lack of knowledge of the organisms dwelling in soils and sediments and of their roles in ecological processes constitutes a major barrier to understanding how ecosystems operate. This was the overriding conclusion of the series of thirteen Biodiversity and Ecosystem Function workshops held during the first phase of the IUBS (International Union of Biological Sciences) / SCOPE (Scientific Committee on Problems in the Environment) / UNESCO (United Nations Educational and Scientific Organization) DIVERSITAS programme in 1988-94, and is reflected in the extensively peer-reviewed *Global Biodiversity Assessment* co-ordinated by UNEP (United Nations Environment Programme) with support of the GEF (Global Environment Facility) and published in November 1995. The topic is not merely an academic pursuit. In order to forecast the implications of perturbations in a site for both the species present and ecosystem functioning, whether direct (e.g. pollution, physical disturbance) or indirect (e.g. climate change), methods need to be available that can be used to ascertain what organisms are present in a sample.

As the next step towards addressing this concern with the second phase of DIVERSITAS, SCOPE decided to establish a series of workshops to review our existing knowledge on soils and sediments at its General Assembly in Tokyo in 1995. At the same time, the IUBS and UNESCO agreed to collaborate in the complementary task of examining existing methods of study, with a view to developing and testing an integrated protocol and manuals to enable soil and sediment biodiversity to be studied in comparable ways. Input on microbiological groups is being co-ordinated through the IUBS/IUMS (International Union of Microbiological Societies) Committee on Microbial Diversity, established in 1994. The first step in the process has been to produce an authoritative manual of the techniques now used for different groups of organisms. This will form the basis for discussions within DIVERSITAS on the construction of the integrated synthetic sampling protocol, but as that will require extensive field-testing after development before it can be commended for general use, I am pleased to see this compilation of methods issued now.

In concluding his autobiography, *The Naturalist*, published in 1994, the entomologist Edward O. Wilson, one of the pioneers in alerting the world to the biodiversity crisis, points to the microscopic world of the soil as the area with the most exciting challenges today, and which he would embark upon if starting anew. If this book stimulates ecologists and other biologists to more thoroughly explore the almost incomprehensible array of organisms in soil and sediment samples and elucidate their ecological role and interdependencies, then it will have served its purpose.

David L. Hawksworth
President, IUBS

International Mycological Institute
Egham
Surrey TW20 9TY

17 April 1996

Preface

The functioning of soil is vital for the survival of the biosphere in its present form, but soil is also one of the most poorly researched habitats on Earth. Microbes and invertebrates maintain soil structure, soil fertility and mediate important ecosystem processes such as decomposition and mineral cycling. In many regions of the world, rapidly expanding human populations have destroyed the soil physico-chemical environment through inputs of chemicals from the atmosphere, disposal of waste products, ground water contamination and physical modification by cultivation and erosion. Information on the effects of these changes on the diversity of soil organisms, on the alteration of key mineral cycling functions (carbon, nitrogen, sulphur, phosphorus, potassium and oxygen) and on the maintenance of plant diversity in the biosphere is fragmentary at best. Global warming research has revealed the key role of the soil biota in regulating methane, nitrous oxide and carbon dioxide losses from the soil, which affect atmospheric and aquatic systems. However, ecological principles derived from the study of macrobiotic communities have been applied uncritically to communities of soil organisms and provide only an untested and approximate basis for predicting the sustainability of soils. The soil biota is also an important source of organisms with biologically active properties for industrial and pharmaceutical intent, for the biological control of agricultural pests and for the remediation of pollutants, hazardous wastes and degraded environments. Soil microbes have a great impact on human life and activities and have enormous potential for providing economic benefits.

Equally, the biotas inhabiting freshwater and marine sediments play critical roles in the decomposition of aquatic and terrestrial materials. Their role in the cycling of nutrients is also vital to the productivity of seas and lakes, which are harvested worldwide by mankind. Increasingly, however, pollution of the marine environment by oil and other industrial and human wastes, the dumping of hazardous wastes (including radionuclides) at sea and the depletion of fish stocks is forcing Governments to re-examine their policies concerning the marine environment. Some progress has been made since the environmental catastrophes such as the eutrophication of Lake Eyrie in the 1960s, but it has not been in proportion to increased demand by the human population. The growth in fish and shellfish farming and the destruction of coastal mangrove forests in tropical countries is adding additional pressure to ensure that marine and freshwater food stocks can be managed sustainably. However, complex issues such as decline in fish production cannot be addressed until benthic productivity and the roles organisms play in decomposition and nutrient recycling are elucidated.

Although thousands of species inhabit one cubic metre of temperate soil or sediment, the identity of most of these organisms and their contribution to sustaining the biosphere are largely unknown. As biotic diversity, in most cases, increases from the poles to the tropics, the biotas of which are less well recorded, the numbers of species will be larger and their identities even more of a mystery. The measurement of species diversity in soil coupled with sustainability assessments of ecosystem processes in soil is increasingly being considered to be a high priority in global biodiversity efforts. Although efforts at the global level have consistently highlighted the need to study soil organisms, there are few scientists with systematic or ecological expertise in this group. Soil is a convenient and easily sampled medium for biodiversity analysis with potential for the development of objective recording techniques.

Now, after two decades of decline, there is suddenly renewed interest in soil biology on an unprecedented level. The origins of this interests are many and various: natural product discovery, biodiversity estimates, soil management for sustainable agriculture, alleviation of the effects of erosion, remediation of pollution, investigation of land use change and fertility improvement by environmentally friendly means. There is also a more widespread realization that, although soil organisms are a key component of ecosystem processes, the identity of only a few is known and their functional diversity and its role in ecosystem processes are often obscure.

In order to meet the above challenges and capitalize on the biodiversity resources of the soil, much synthesis of existing work and application of new techniques will be required. This book is a step towards that goal by ensuring that as many methods for the examination of organismal diversity as possible are available in one volume, instead of being scattered throughout microbiological, zoological and botanical literature. In addition, these methods are focused on soils and sediments rather than being general methods. However, it is realized that in any sampling scheme it may not be feasible to sample all groups, and it may not be a study objective to do this, and so each chapter is self-contained and may be 'extracted' whole for use in any particular survey. This has resulted in some duplication of the methods. In defence, it must be pointed out that often so-called 'general methods' have been found to have (sometimes serious) limitations when applied to specific groups, and often need modification in the light of experience. Indeed, this has been the experience of contributors to this book.

As the focus of the book is organismal biodiversity, a broadly organismal arrangement has been adopted. The contributions on the different groups are all by recognized specialists, each of whom has an intimate working knowledge of their particular group, but each was asked to address a series of headings in their chapters. Quantification is given wherever possible, but is not feasible in some microbial groups for reasons of habit, inadequate species concepts or lack of techniques.

This compilation of methods is being made available now because of the demand for information on techniques, but it is also a step towards a general synthetic protocol, as described in the Foreword.

Geoffrey S. Hall
Editor
January 1996

Acknowledgements

This work would not have been realized without the initiative of Dr Pierre Lasserre (Director, Division of Ecological Sciences, UNESCO) in developing and supporting this project as a component of UNESCO's contribution to the DIVERSITAS programme.

The camera-ready copy was finalized for production by Ms Marilyn S. Rainbow of the International Mycological Institute.

Heterotrophic Bacteria: the Cultivation Approach

1

JAMES T. STALEY

Department of Microbiology, University of Washington, Seattle, WA 98195, USA.

The Importance of Heterotrophic Bacteria in Ecosystem Processes

The major role of heterotrophic bacteria in ecosystems is the degradation of organic matter. This process is referred to as decomposition. If the organic matter is completely converted to inorganic matter, including carbon dioxide, phosphate, ammonia and sulphate, the process is referred to as mineralization. The mineralization of organic matter, which is derived from primary producers, results in its being recycled, so that these substances are again available for primary producers. Fungi and protozoa are also involved in recycling activities in natural ecosystems. (See this volume, Chapters 7, 9-12).

Most of the twelve or so phylogenetic groups of the *Eubacteria* contain heterotrophic members, which indicates the genetic diversity of heterotrophic bacteria. This genetic diversity mirrors the functional diversity of heterotrophic bacteria as well. For example, although animals carry enzymes for fermentation of glucose via the glycolytic pathway, this is only one of a large variety of fermentation pathways found in facultative aerobic and anaerobic heterotrophic bacteria. These other pathways include mixed acid, propionate, acetone/butanol, amino acid and purine fermentations, which result in the degradation of all sorts of organic substrates, e.g. cellulose, starch, proteins and amino acids to list a few.

Aerobic heterotrophic bacteria are equally versatile and important in the degradation of organic matter in natural ecosystems. The combined activities of the aerobes and anaerobes, which often live in close association in soils and sediments, result in the ability of the bacterial population to degrade a myriad assortment of naturally occurring compounds, as well as toxic xenobiotic compounds such as PCBs and pesticides.

Heterotrophic bacteria are widely distributed in soils and sediments throughout the Earth. They are remarkable as a group in growing over wide ranges of pH, salinity, temperature, redox potential and substrate concentrations. For this reason, they grow in some of the most extreme habitats available for life on Earth, which are not always occupied by other decomposers.

Approaches to the Study of the Diversity of Heterotrophic Bacteria

Two separate approaches have been used to assess the diversity of heterotrophic bacteria in natural communities. The traditional approach requires that the organisms are cultivated in pure culture and characterized phenotypically and genotypically. More recent molecular approaches do not require that bacteria be cultivated; instead, the community diversity is assessed by an examination of its extracted nucleic acids, in particular, DNA. Both approaches have advantages and disadvantages. The principal advantage of using the classical cultivation approach is that organisms are isolated and therefore available for further study. For example, if one wishes to conduct taxonomic studies of strains from a habitat, it is essential to have the strains in culture so that they can be compared head-on with other strains from the community and known type strains from culture collections. Also, if the microbiologist is interested in the discovery of new antimicrobial agents, enzymes or other biotechnologically important features, it is desirable, if not essential, to have the organism in culture.

The primary disadvantage of the cultivation approach is that the most numerous bacteria from many natural communities cannot be grown in pure culture using current procedures. Indeed, typically less than 1% of the total bacteria are recovered by cultivation techniques from most natural soil and aquatic habitats. In contrast, molecular approaches do not require that the bacteria be isolated. However, there are disadvantages to the molecular approach too. Some disadvantages of using molecular approaches include the difficulty in lysing all bacteria from natural communities, the presence of DNA from phages and higher organisms in the community, the extraction of DNA from dead bacteria, and the difficulty in quantification of important species from the habitat, to mention a few. Furthermore, it is often not possible to determine the physiological type or species from its 16S rDNA sequence. For these reasons, it is impossible to determine diversity indices and species diversity of heterotrophic bacteria accurately in most communities using either cultivation or molecular approaches. Ideally, both approaches should be used at the same time so that a comparison can be made between the phylogenetic types from the culture collection and from the extracted DNA.

In this section, only the cultivation approach is discussed, including both general and specific techniques. Particular emphasis is given to the cultivation of aerobic and facultative aerobic bacteria in this section. Selective media and growth conditions are needed to study special heterotrophic groups such as sulphate reducers, iron and manganese reducers, methylotrophic bacteria, denitrifiers, and others (see this volume, Chapter 2).

Viable Recovery of Heterotrophic Bacteria from Soil and Sediment Communities

In order to evaluate the ability of a procedure to recover viable bacteria from a community, it is essential to be able to conduct a census of all the indigenous bacteria. This is accomplished by total microscopic counting.

Total Microscopic Count Using Fluorescent Dyes

The most commonly practised microscopic counting techniques use fluorescent dyes such as acridine orange (Hobbie *et al.*, 1977) and DAPI (4',6-diamidino-2-phenylindole; Porter and Feig, 1980). These dyes form complexes with the intracellular nucleic acids, thereby rendering the cell fluorescent. After fixation and staining the cells are concentrated on a 'Nucleopore' filter (0.2 μm pore size; Nucleopore Co.) that has been pre-stained with a non-fluorescent black dye such as irgalan black. The cells will appear orange or green in colour when observed by epifluorescence microscopy using an appropriate halogen or ultraviolet light source. It is essential to use appropriate controls in these staining procedures. For example, chlorophyll *a* has a natural fluorescence, and so cyanobacteria and algae will fluoresce even without staining. To correct for this, it is important to use control filters with unstained samples. As soil and sediment samples contain minerals and organic detritus which may fluoresce, it is important to compare unstained samples with stained samples in order to identify bacteria. Finally, particles may have bacteria adhering to their under surfaces - these bacteria will not be detected because the light cannot be transmitted through the particles.

Viable Enumeration

Although, as mentioned previously, traditional viable enumeration procedures have their drawbacks, they can be useful in assessing microbial diversity. In some habitats, a close correspondence occurs between total and viable counts (Staley, 1980). In such habitats, there is considerable value in assessing the diversity on cultivation media. Recently, Schut *et al.* (1993) grew almost all of the bacteria in the marine water column using dilute media and prolonged incubation. More research is needed using this approach to determine if such procedures are commonly applicable to soil and sediment communities. If so, they will revolutionize the assessment of the diversity of heterotrophic bacteria. However, even if all the bacteria cannot be cultivated, there is considerable value in assessing the diversity of the bacteria that can be grown from natural communities. Furthermore, the diversity of those species that can be cultivated is of interest because they are members of the community. Moreover, a medium that will grow a given species, should be able to grow the same species from any location in the same habitat, or from another habitat in which it occurs. Thus, cultivation procedures can be used to inventory those species that can be grown such as *Caulobacter* spp. in the water column of lakes (Staley *et al.*, 1987). In the following section, some recommended viable counting approaches are described for growth of aerobic and facultative aerobic heterotrophic bacteria. Other media and approaches should be used for the growth of particular anaerobic groups such as the denitrifiers or sulphate reducers.

Spread Plate Procedure

One of the simplest viable counting procedures for heterotrophic bacteria is the spread plating technique. A medium is prepared and a gelling agent such as agar is added. Plates are poured and dilutions from the habitat are spread on the medium using a glass rod bent in the shape of a hockey stick. The principal advantage this procedure has over pour plating is that bacteria from many communities, especially sediment habitats, would be killed by exposure to the high temperatures encountered when samples are mixed in molten pour plate media. A major question facing microbiologists interested in viable counting is the choice of media since there is no one 'universal medium' that can be used to cultivate all heterotrophic bacteria. One approach is to use a medium that is derived from the natural habitat itself. For example, a medium can be made from the soil or sediment of the habitat. Soil extract media are commonly used for this purpose. Soil extract can be used directly and solidified with agar, or it can be amended with organic or inorganic media and reagents. Alternatively, a general purpose medium can be used which would allow the growth of a variety of different bacteria. It is recommended that both approaches be used, at least until one medium is determined to be better than another, or it is found that a combination of media allows the growth of different strains.

One general purpose medium on which many heterotrophic bacteria grow is PYGV medium which contains low concentrations of peptone, yeast extract, glucose, and a vitamin and salts solution. For marine sediments, half-strength or full-strength artificial seawater (ASW) can be added. Another medium that can be used in marine environments is Ordal's cytophaga agar, which contains acetic acid as the major carbon source. Anaerobic bacteria such as many *Clostridium* spp. will grow on these media if the plates are incubated under anaerobic conditions. Other complex media may be tried as well. In addition, selective media can be used to allow for the growth of specific physiological or nutritional groups of heterotrophs such as sulphate reducers that would not grow using the media described above.

Unwanted fungal growth may pose difficulties for assessing bacterial growth particularly from soil and possibly freshwater sediment samples. To reduce this, cycloheximide can be added to the medium to inhibit growth, usually at about 10 $\mu g\ ml^{-1}$. The samples to be plated should be processed soon after collection. They should be kept in the dark and refrigerated to keep them at, or below, the ambient temperature (this is especially important for the sediment bacteria, which are typically never exposed to high temperatures). Likewise, for processing in the laboratory, temperatures should be kept low (ideally less than 15°C for temperate zone sediments). Finally, incubation temperatures should correspond to those of the environments from which the samples were collected. Sediment and soil samples should not be incubated at temperatures higher than room temperature.

MPN Enumeration

Extinction dilution procedures using the most probable number (MPN) technique are also useful in determining the viable count. This is because this procedure provides a completely aqueous environment of high water activity in which to grow the bacteria. Soil and sediment samples are mixed in a sterile non-growth buffer for quantitative serial dilution. Replicates of each dilution are transferred to a liquid growth medium and incubated under appropriate conditions. The highest dilution tubes that become turbid during the course of incubation are scored as positive for viable bacteria. An MPN enumeration table (in *Standard Methods for the Examination of Water and Wastewater*, 1980) can be used to determine numbers. The choice of media will depend on the type of organism desired. Inasmuch as each of these tubes is an enrichment culture, selective conditions can be established for the isolation of specific physiological groups.

A disadvantage of this procedure is that the organisms are still not isolated. Furthermore, almost all of the tubes contain several bacteria from the initial inoculation so there is a succession of types occurring as incubation proceeds. Nonetheless, this procedure provides an alternative approach to plating for the enumeration of viable bacteria.

The results of Schut *et al.* (1993) and, more recently, Bianchi and Giuliano (1996) suggest that it is possible to grow many of the most numerous bacteria in sterile water from the environment in MPN extinction dilution tubes after prolonged incubation from two to six months. Therefore, this procedure is recommended for both sediment and soil habitats, as it may provide a means for accurately assessing the most numerous bacteria in the habitat, and also lead to their cultivation and phenotypic and phylogenetic characterization.

Assessing Recoverability and Estimating Diversity

After the colonies have developed on the plates, they should be counted. The recoverability for the medium is determined by dividing the viable count by the total microscopic count. If the recoverability ratio is 1% or less, then the habitat is regarded as an oligotrophic or mesotrophic habitat (Staley *et al.*, 1982). If several media have been used, it is of interest to compare the recoverability of one with the other(s). Furthermore, it would be of interest to assess whether different species are being grown using different media. This could be assessed by replica plating the colonies from one medium to the other. From this approach it would be possible to determine how many cultivable species there are from the habitat by summing the numbers of those species that cannot be grown on the other media being tested with those that can be grown on more than one medium. To my knowledge, however, this approach has not been used experimentally.

A diversity index, such as the Shannon-Weaver Index (Shannon and Weaver, 1949) can be used to determine the diversity of strains based on colony type. Thus, if this is of interest, it is necessary to describe and characterize each of the colony types and count their number. Even if a diversity index is not determined, it is useful to have an estimate of the number of each of the strains based on its colony morphology.

Phylogenetic Diversity of the Cultures Collected

The colonies on the spread plates described above comprise the collection of bacteria from the habitat of interest. Representative colony types are then picked from plates and they are purified by re-streaking. Alternatively colonies can be randomly picked. Using either of these approaches, collections of one hundred or more strains can be quickly assembled. The next step is to organize the strains into groups before characterizing representative strains for phenotypic and genotypic characteristics.

Grouping of strains

Because it is likely that many of the strains are similar to other strains from the collection, especially if they are randomly selected from the plates, it is helpful to have some test(s) for grouping strains into similar types. This is desirable because it reduces the number of strains that need to be characterized to a manageable number. These representative strains can then be subjected to more intensive characterization.

Whole cell fatty acid analysis

This technique is very useful for this purpose. Most bacteria from natural environments have not been studied and so their fatty acid profiles are not found in the data banks of commercial firms such as the MIDI data bank (Microbial Identification Inc., Delaware). This is because the commercial firms are most concerned with the identification of pathogenic or potentially pathogenic bacteria. Furthermore, many of these environmental strains will not grow on the test media used for characterization of type strains. Nevertheless, it is possible to determine the fatty acid composition of a strain under known growth conditions. Furthermore, it is possible to construct a dendrogram that groups the strains based on the similarity of their fatty acid profiles using cluster analysis programmes.

Using this information, it is possible to assess which strains are very closely related to other strains and which are very different. Thus, if the original collection had one hundred strains, perhaps five or fifteen different subgroups can be identified. A representative member of each of these subgroups can then be subjected to further analysis using phenotypic and genotypic features. In this manner, the numbers of strains that need to be intensively studied are reduced. Only the representative type strain of each group need be more fully characterized, as described below.

Phylogeny of representative strains

To assess the phylogeny of a strain, it is necessary to determine its 16S rDNA sequence. This is very valuable for identification of new bacteria, particularly those strains that are aerobic and Gram-negative. Phenotypic features are frequently insufficient to allow for a determination of the identification of an environmental isolate to the genus and species level. The procedure involved in sequencing includes the following steps (after Woese, 1987; Dykesterhouse *et al.*, 1995):

- Isolate genomic DNA from strain.
- Amplify 16S rDNA by PCR using universal primers.
- Ligate 16S rDNA gene into sequencing vector.
- Conduct sequence comparison with database, such as the 'Ribosome Database Project' at the University of Illinois.
- Subject sequence to phylogenetic analysis using maximum likelihood, parsimony and distance matrix methods.

The results of this analysis will indicate whether the strain is a new organism and will further indicate what its closest relatives are. By knowing its closest relatives it is possible to conduct additional tests to assess whether a new species or genus has been found (see also this volume, Chapter 5).

Taxonomy of new species

One of the primary advantages of the culture approach is that it allows for the determination of whether the newly isolated soil or sediment strains are new species or members of known species. For example, if they are strains of known species the 16S rDNA sequence should be very similar, if not identical, to that of the known type strain. Furthermore, the whole cell fatty acid analysis procedure described above will prove to be very useful in confirming their identification in that they can be compared head-on with the type strain. The same is true of other phenotypic and genotypic features. However, little work has been conducted on taxa from some environments such as marine habitats. So new taxa would be expected to be found in these habitats.

Description of new taxa

Studies indicate that strains that show greater than 3% difference in homology between 16S rDNA sequences compared with known strains comprise new species (Stackebrandt and Goebel, 1994). However, it is important to conduct a variety of phenotypic and genotypic tests to permit the two species to be distinguished from one another. On the other hand, if the 16S rDNA sequence similarity is greater than 97%, it is essential to conduct DNA/DNA hybridization experiments to determine whether the strain is a new species (Wayne *et al.*, 1987). Conditions for DNA/DNA hybridization are described by Johnson (1994).

If the results from DNA/DNA hybridization tests indicate that this is a new species, phenotypic and genotypic features need to be identified to determine how the new species differs from known species. Some of the types of phenotypic and genotypic properties that are useful for this purpose are described below.

Phenotypic features

The phenotypic features of interest include morphological, metabolic and physiological features. Morphological tests include cell shape and size, motility, Gram's stain reaction, presence of endospores, presence of appendages and gas vacuoles, etc. (Smibert and Krieg, 1994). Nutritional and physiological features to test include carbon source requirements, vitamin requirements, aerobic or anaerobic growth, sugar fermentation, use of alternate electron acceptors (e.g. nitrate and sulphate), pH range for growth, temperature range for growth, and a variety of other tests depending on the group to which it may belong. Note that it is important to include the closest known related strains for comparison in these tests. These should always be the known type strains of the species.

Genotypic features

In addition to the DNA/DNA hybridization tests already described, it is important to determine the DNA base composition of the strains in question. This can be determined using purified DNA by several techniques such as thermal denaturation or HPLC analysis of purines and pyrimidines from hydrolysed DNA (Smibert and Krieg, 1994).

Spatial and Seasonal Distributions of Species in the Environment

If a strain from a community is of particular interest, it is possible to determine its numerical importance in the community. This would allow for an assessment of its distribution in the habitat both on a spatial and temporal basis. Thus, for example, it would be possible to determine how numerous it is at different locations in the soil or sediment profile and, further, whether it is more common during one season of the year or another. This type of question can be answered for pure cultures using either the fluorescent antibody (FA) approach or by use of fluorescent nucleic acid probes (Amann *et al.*, 1995).

Bibliography

Amman, R.I., Ludwig, W. and Schleifer, K.H. (1995) Phylogenetic identification and *in situ* detection of individual microbial cells without cultivation. *Microbiological Reviews* 59, 143-169.

Bianchi, A. and Giuliano, L. (1996) Enumeration of viable bacteria in the marine pelagic environment. *Applied and Environmental Microbiology* 62, 174-177.

Dykesterhouse, S.E., Gray, J.P., Herwig, R.P., Lara, J.C. and Staley, J.T. (1995) *Cycloclasticus pugetii* gen. nov., sp. nov., an aromatic hydrogen-degrading bacterium from marine sediments. *International Journal of Systematic Bacteriology* 45, 116-123.

Hobbie, J.E., Daley, R.J. and Jaspar, S. (1977) Use of Nucleopore filters for counting bacteria by fluorescence microscopy. *Applied and Environmental Microbiology* 33, 1225-1228.

Johnson, J.L. (1994) Similarity analysis of DNAs. In: Gerhardt, P. (ed.) *Methods for General and Molecular Bacteriology*. American Society for Microbiology, Washington DC, pp. 655-700.

Porter, K. and Feig, Y.S. (1980) The use of DAPI for identifying and counting microflora. *Limnology and Oceanography* 25, 943-951.

Schut, F., de Vries, E.J., Gottschall, J.C., Robertson, B.R., Harder, W., Prins, R.A. and Button, D.K. (1993) Isolation of typical marine bacteria by dilution culture: growth, maintenance and characteristics of isolates under laboratory conditions. *Applied and Environmental Microbiology* 59, 2150-2160.

Shannon, G.E. and Weaver, W. (1949) *The Mathematical Theory of Communication*. University of Illinois Press, Urbana.

Smibert, R.M. and Krieg, N.R. (1994) Phenotypic characterization. In: Gerhardt, P. (ed.) *Methods for General and Molecular Bacteriology*. American Society for Microbiology, Washington DC, pp. 607-654.

Stackebrandt, E. and Goebel, B.M. (1994) Taxonomic note: a place for DNA-DNA reassociation and 16s rRNA sequence analysis in the present species definition in bacteriology. *International Journal of Systematic Bacteriology* 44, 846-849.

Staley, J.T. (1968) *Prosthecomicrobium* and *Ancalomicrobium*: new prosthecate freshwater bacteria. *Journal of Bacteriology* 95, 1921-1942.

Staley, J.T (1980) Diversity of aquatic heterotrophic bacterial communities. In: Schlessinger, D.C.A (ed.) *Microbiology*. American Society for Microbiology, Washington DC, pp. 321-322.

Staley, J.T., Lehmicke, L.G., Palmer, F.E., Peet, R.W. and Wissmar, R.C. (1982) Impact of Mount St Helens eruption on bacteriology of lakes in the blast zone. *Applied and Environmental Microbiology* 43, 664-670.

Staley, J.T., Konopka, A.E. and Dalmasso, J. (1987) Spatial and temporal distribution of *Caulobacter* spp. in two mesotrophic lakes. *FEMS Microbiology Ecology* 45, 1.

Wayne, L.G., Brenner, D.J., Colwell, R.R., Grimont, P.A.D., Kandler, O., Krichevsky, M.I., Moore, L.H., Moore, W.E.C., Murray, R.G.E, Stackebrandt, E., Starr, M.P. and Trüper, H.G. (1987) Report of the ad hoc committee on reconciliation of approaches to bacterial systematics. *International Journal of Systematic Bacteriology* 37, 463-464.

Woese, C. (1987) Bacterial evolution. *Microbiological Reviews* 51, 221-271.

Chemolithotrophic Bacteria 2

Sheridan K. Haack

US Geological Survey, Water Resources Division, 6520 Mercantile Way, Suite 5, Lansing, MI 48911, USA.

The chemolithotrophic bacteria are capable of deriving energy for cell metabolism from inorganic reactions in the absence of light (Kelly, 1992). Some chemolithotrophic bacteria use reduced inorganic compounds as an electron donor for cell metabolism, which results in the oxidation of these inorganic compounds. Other chemolithotrophic bacteria use oxidized inorganic compounds as an electron sink and carry out the reduction of these compounds. Many chemolithotrophic bacteria are capable of autotrophy, the process by which cellular carbon is derived strictly from CO_2. Such bacteria are termed chemolithoautotrophs. This report will consider primarily those bacterial groups which contain members capable of chemolithoautotrophic growth and which carry out anaerobic transformations of CO_2 or oxidations and reductions of N-, S- and Fe-containing compounds. A summary of representative chemolithoautotrophic bacteria is presented in Table 2.1, organized first by the reactions catalyzed, and second by phylogenetic criteria.

The capacity for chemolithoautotrophic metabolism is widespread among the phylogenetic domains *Bacteria* and *Archaea* (as defined by sequence analysis of ribosomal RNAs (rRNAs); Woese *et al.*, 1990; Olsen *et al.*, 1994). As shown in Table 2.1, some types of chemolithoautotrophic metabolism (such as the ability to oxidize NH_4) are restricted to a few highly related (eu)bacterial genera. Similarly, the chemolithoautotrophic reaction by which H_2 is used to reduce CO_2 to CH_4 (methanogenesis) is restricted to a highly related group of genera in the phylogenetic domain *Archaea*. However, other chemolithoautotrophic reactions (such as the oxidation of sulphur compounds) are carried out by numerous (eu)bacterial and archaeal genera which exhibit few or no additional phenotypic or phylogenetic relationships.

Table 2.1. Chemolithotrophic bacteria involved in N, S, Fe and CH_4 transformations

These lists identify described bacteria capable of carrying out the indicated reaction under chemolithoautotrophic conditions. In some cases, bacteria in the same or related genera may carry out the same reaction under chemolithoheterotrophic conditions. In other cases, bacteria in related or unrelated genera may carry out similar reactions under strictly heterotrophic conditions. Therefore, the list does not include all bacteria known to carry out each transformation. Most groups are described in Balows *et al.* (1992) or Staley *et al.* (1989), unless a specific citation indicates otherwise.

Bacteria Involved in Nitrogen Transformations

Oxidizers/Nitrifiers[a]
Oxidize NH_3 to NO_2^-
(Eu-)Bacteria: Proteobacteria, β subclass
Nitrosococcus (except *N. oceanus*)
Nitrosolobus
Nitrosomonas
Nitrosospira
Nitrosovibrio
(Eu-)Bacteria: Proteobacteria, γ subclass
Nitrosococcus oceanus

Oxidize NO_2^- to NO_3^-
(Eu-)Bacteria: Proteobacteria, α subclass
Nitrobacter
Nitrococcus
Nitrospina
Nitrospira

Reducers/Denitrifiers[b]
(Eu-)Bacteria: Proteobacteria, α subclass
Paracoccus denitrificans
(Eu-)Bacteria: Proteobacteria, β subclass
Alcaligenes eutrophus
Thiobacillus denitrificans
(Eu-)Bacteria: Proteobacteria, γ subclass
Thiomicrospira denitrificans
Uncertain affiliation
Thiosphaera pantotropha

Bacteria Involved in Iron Transformations

Oxidizers[b]
Uncertain affiliation
Gallionella
Leptospirillum ferrooxidans
Sulfobacillus
(Eu-)Bacteria: Proteobacteria, β subclass
Thiobacillus ferrooxidans

Reducers[b]
(Eu-)Bacteria: Proteobacteria, γ subclass
Shewanella putrefaciens
(Eu-)Bacteria: Proteobacteria, δ subclass
Geobacter sulphurreducens[c]
Uncertain affiliation
'*Pseudomonas*' sp.[d]
Strain BrY[e]

Bacteria Involved in Sulphur Transformations

Oxidizers[b]
(Eu-)Bacteria: Proteobacteria, α and β subclasses
Paracoccus denitrificans
Thiobacillus (some species)
(Eu-)Bacteria: Proteobacteria, γ subclass[f]
Beggiatoa (some species)
Thiomicrospira
Thiothrix (some species)
(Eu-)Bacteria: Proteobacteria, ε subclass
Thiovulum (some species)
Archaea: Sulfolobales
Acidianus
Desulphurolobus
Sulfolobus
Uncertain affiliation
Leptospirillum ferrooxidans
Thermothrix thiopara
Thiosphaera pantotropha

Reducers[b]
(Eu-)Bacteria: Proteobacteria, δ subclass
Desulfobacter hydrogenophilus
Desulfobacterium
Desulfonema limicola
Desulfosarcina variabilis
Desulfotomaculum orientis
Archaea: Thermoproteales
Pyrobaculum (some species)
Pyrodictium (some species)
Thermoproteus (some species)
Archaea: Sulfolobales
Acidianus
Desulphurolobus

Bacteria which Convert H_2 and CO_2 to Acetate or Methane

Acetogens[b]
(Eu-)Bacteria: Gram +ve bacteria; Subdivision A2
Acetobacterium
Clostridium (some species)
Eubacterium limosum
Peptostreptococcus productus
(Eu-)Bacteria: Gram +ve bacteria: Subdivision A3
Sporomusa

Methanogens (All *Archaea*)[a]
Methanobacteriaceae (3 genera)
Methanococcaceae (1 genus)
Methanocorpusculaceae (1 genus)
Methanomicrobiaceae (6 genera)
Methanosarcinaceae (7 genera)
Methanothermaceae (1 genus)

a Phylogenetically coherent groups based on 16S rRNA sequences and/or physiology.
b Phylogenetically diverse groups.
c Caccavo *et al.* (1994)
d Balashova and Zavarzin (1980)
e Caccavo *et al.* (1992)
f Group in which sulphur-oxidizing chemolithotrophic symbionts of marine invertebrates have been placed by 16S rRNA sequences.

Chemolithoheterotrophic bacteria derive their cellular carbon from organic compounds while continuing to carry out oxidations or reductions of inorganic compounds. Some of the chemolithoautotrophic bacteria listed in Table 2.1 are also facultative chemolithoheterotrophs. For example, some *Thiobacillus* species grow on either CO_2 or a variety of organic acids and sugars while oxidizing reduced sulphur compounds (Kelly, 1992). In contrast, *Geobacter sulphurreducens* can grow using either CO_2, acetate or formate coupled to the reduction of Fe(III) (Caccavo *et al.*, 1994). Several groups of bacteria best known for heterotrophic metabolism (in which both energy and cellular carbon are derived from organic compounds) contain members which may also derive energy from chemolithoheterotrophic reactions. Chemolithoheterotrophs are especially important in reductive reactions involving N, S, Fe, Mn and trace element (As, Cr, Cu, Se, U) transformations (Kelly, 1992; Nealson and Myers, 1992; Lovley, 1995).

Importance in Ecosystem Processes

The ecological significance of chemolithotrophic bacteria, and the reactions they catalyze, is immense. Chemolithotrophic reactions affect the abundance, distribution and form of such major elements as C, N, S, Fe and Mn and are integral to the biogeochemical cycles of these elements. Through their effects on soil or sediment mineralogy and water chemistry, chemolithotrophs exert considerable control over many ecosystem processes. In addition, evidence from sulphide-rich marine environments suggests that chemolithotrophs may directly affect patterns of macrofaunal diversity through symbiotic relationships.

Chemolithotrophic reactions result in the oxidation and reduction of essential nutrients such as nitrogen, iron and sulphur. Significant segments of the nitrogen and sulphur biogeochemical cycles depend on the metabolism of chemolithotrophs (Brock and Madigan, 1988; Paul and Clark, 1989). Bacterial conversions of nitrogen, sulphur and iron result in transport of nutrients from one ecosystem to another on local, regional and global scales, through the production of gases (e.g. N_2, H_2S) or of forms easily leached from soils to water (NO_3^-, Fe^{2+}). The productivity of many ecosystems may be based on the success of soil or sediment chemolithotrophs in providing these essential elements in the appropriate forms for uptake by plants. In addition, the chemolithotrophic reduction of CO_2 to CH_4 is an important component of the carbon cycle and results in the production of a 'greenhouse' gas significant in global climate change.

In sediments and ground water, oxidized forms of N, Mn, Fe, S and ultimately CO_2 form a sequence of 'terminal electron acceptors' for microbial growth (Lovley, 1991; Nealson and Myers, 1992; Chapelle, 1993; Steenbergen *et al.*, 1993). The utilization of these electron acceptors by bacteria results in predictable oxidation/reduction (redox) gradients, which are similar whether they occur on the scale of millimetres (wetland sediments) or kilometres (ground water flowpaths). These gradients, through their effects on sediment or ground water quality, define ecotones across which macroscopic flora and fauna change, and may govern the suitability of certain habitats for other organisms.

Chemolithotrophic bacteria carry out reactions which affect the precipitation or dissolution of minerals. Specific chemolithotrophic reactions have been implicated in such diverse phenomena as acid-mine drainage (Brock and Madigan, 1988), the formation of magnetite (Lovley, 1991) and the weathering of sandstones in building materials (Meincke *et al.*, 1989). Chemolithotrophic reactions also affect the release (through reduction of oxidized iron-containing minerals) of phosphate and trace metals from aquatic sediments or, conversely, the precipitation of trace metals from solution with microbially produced sulphides (Lovley, 1991). Chemolithotrophic reactions affect the form and speciation of trace metals such as As, Cr, Cu, Se and U and in doing so affect their relative toxicity to other organisms (Ahmann *et al.*, 1994; Laverman *et al.*, 1995; Lovley, 1995).

Finally, chemolithotrophic sulphur-oxidizing endosymbionts appear to be the sole source of energy for a variety of marine animals in the phyla *Annelida, Bivalva, Pogonophora* and *Vestimentifera* (Fisher, 1990; Cavanaugh, 1994). Very recently, chemolithotrophic, ectosymbiotic, sulphur-oxidizing bacteria were identified (based on 16S rRNA analyses - none have been isolated) in association with a marine nematode (Polz *et al.*, 1994). In both cases, the host organisms have been found in proximity to deep-sea hydrothermal vents, or in other marine environments where reduced sulphur compounds are abundant. Although there is currently no terrestrial or freshwater example of this unique symbiosis, its significance to microbial and macrofaunal diversity is immense, and there is no reason to expect that additional unique symbioses might not be found in terrestrial or freshwater habitats.

Sampling/Extraction Techniques

Most isolations of bacteria are based on enrichment culture. Enrichment culture is a process which encourages the growth of selected bacteria based on their metabolic specificity. In theory, chemolithotrophic growth can be encouraged by providing a growth medium which contains only CO_2 plus the oxidant and reductant required to generate the energy-yielding reaction. For example, NH_4-oxidizers would be enriched from an environmental sample in a simple medium which supplied NH_4, O_2, CO_2 and basic nutrients such as P and S (and perhaps vitamins) in a suitable form. Such a medium should theoretically be suitable only for those bacteria capable of chemolithotrophic growth through NH_4 oxidation; therefore, all bacteria in the sample which grow by other means would be excluded by this growth medium. One key to successful enrichment is choosing an inoculum source expected to contain large numbers of the type of organism sought. By this reasoning, NH_4-oxidizers would be especially sought in environments where NH_4 was abundant and where there was geochemical evidence of NH_4 oxidation. Numerically dominant NH_4 oxidizers might be selected by diluting the environmental sample and starting the enrichment culture with the most dilute subsample which still exhibited evidence of NH_4 oxidation.

Organisms which grew in the enrichment medium would then be transferred periodically to fresh liquid or solid medium. An advantage of solid media is that colonies representing growth from a single cell may develop. Over time, this process should result in the establishment of a pure culture capable of carrying out the chemolithotrophic reaction of interest. Once a pure culture was obtained, a battery of physiological, chemotaxonomic and molecular (DNA- or RNA-based) tests would provide the information required to assign the new isolate to an existing or new bacterial taxon. The quantitative significance of the new organism in its original environment could eventually be assessed by re-examining samples from that environment using specific phenotypic or genotypic characteristics of the isolate. Genotypic methods might include hybridization of nucleic acid probes to samples prepared for microscopy or to whole community DNA or RNA extracted from the sample. Phenotypic methods might include those based on immunological detection, or chemotaxonomic methods based on cell components unique to the isolate (e.g. fatty acids or specific proteins). If these methods were applied to dilutions of the original sample, or to dilutions of DNA extracted from the sample, the most dilute sample which tested positive for the isolate could give an estimate of abundance.

A great deal of information is available regarding appropriate growth media, selective culture methods and likely habitats for specific chemolithotrophic groups of bacteria. Compendia such as *The Prokaryotes* (Balows *et al.*, 1992) and *Bergey's Manual of Systematic Bacteriology* (Staley *et al.*, 1989) provide summary chapters on each of the major bacterial groups, including all the chemolithoautotrophic groups listed in Table 2.1. These summaries identify the habitats and ecology of each group and describe isolation strategies and appropriate growth media. In addition, each summary describes in detail features which distinguish the various taxa. Written by experts, these summaries provide a wealth of practical information, important historical and personal perspectives and a pathway to the detailed literature on each specific group. Other texts, such as *Autotrophic Bacteria* (Schlegel and Bowien, 1989), provide similar information on specific chemolithotrophic groups. In addition, texts such as *Methods for General and Molecular Bacteriology* (Gerhardt *et al.*, 1994) provide summaries of the majority of procedures common to microbiological analyses, and list media and isolation methods for many typical microorganisms, including some chemolithotrophs. Finally, texts such as *The Handbook of Microbiological Media* (Atlas, 1993) list 'recipes' and applications for many of the growth media used in microbiology. Such references provide ideal points of departure for developing an isolation strategy for a particular chemolithotrophic group.

Limitations of Existing Techniques

There are numerous practical impediments to the development of a scheme for the determination of the diversity of specific chemolithotrophic bacteria in soils and sediments. Although enrichment culture methods for isolation of most groups of chemolithotrophic bacteria are well documented, and have successfully yielded the isolates described in Table 2.1, enrichment culture of chemolithotrophs is an exacting process. One major limitation to enrichment culture of chemolithotrophs is the relatively slow growth of isolates and a requirement for complete absence of organic compounds in all medium constituents. A common problem encountered during enrichment culture is overgrowth of the slow-growing chemolithoautotrophs by heterotrophic bacteria utilizing trace levels of organic compounds commonly found in inorganic medium constituents. Time periods of weeks to months to obtain a pure culture are commonly reported. A second limitation is the requirement, in some cases, for growth media which may require low oxygen tension or strictly anaerobic conditions, or the addition of several medium constituents as gases. These limitations are inherent aspects of chemolithotrophic growth and cannot be avoided. In addition, for the majority of bacterial groups, the tests required to assign a new bacterial isolate to a specific taxon require a fully equipped microbiology laboratory and considerable training.

Further impediments to determining chemolithotrophic diversity are associated with the bias inherent in enrichment culture. Methods developed over the last 15-20 years have demonstrated that there are many more bacterial cells (of a variety of metabolic types) in most environmental samples than can be isolated on typical enrichment media. In addition, when it has been possible to re-examine environments with DNA- or RNA-based probes for specific organisms isolated from those samples, it has been noted that the isolated organisms often comprise only a small percentage of the community (Ward *et al.*, 1990). These observations suggest that there is much greater bacterial diversity in nature than is represented in current culture collections, and that the organisms we have cultured may not be the important ones in nature. Some of this bias results from our inability to design appropriate enrichment culture methods for specific physiological groups and from specific choices made by the investigator during the development of an enrichment and isolation protocol. Decisions about the concentrations of growth substrates to be provided or the choice of liquid versus solid culture medium may affect which organism becomes dominant in the enrichment culture. For example, iron-reducing bacteria have only been isolated and studied in pure culture within the last 10 years (Lovley, 1995). Important to the enrichment and isolation of these bacteria was development of a medium which contained iron in a form suitable for their growth (Lovley, 1995). The 'iron-oxidizing' bacteria have proved especially resistant to isolation, although many possess unique morphologies which allow them to be visually detected in environmental samples (Ghiorse and Ehrlich, 1993; Emerson and Revsbech, 1994). In some cases bias may be overcome by use of creative techniques such as gradient culture, which more closely represent the *in-situ* environment (Wimpenny, 1993).

Additional bias results from the selective nature of enrichment culture. For example, recent reports described the isolation of strictly anaerobic methanogenic, acetogenic and sulphate-reducing bacteria from a variety of oxic soils (Peters and Conrad, 1995) and of chemolithoautotrophic nitrifiers from water-saturated (and, hence, anoxic) soils (Both *et al.*, 1992). In each case, these results were not expected since the assumption had been made that the environment precluded the growth of these bacterial groups. In addition, these taxa require specialized culture media and would not be detected in routine procedures for isolating typical (expected) soil bacteria. Therefore, some of the uncultured bacterial diversity in soils may result from the presence of common or well-described taxa which are nevertheless not anticipated in that habitat and which require specialized approaches for their detection.

The wide variety of metabolic strategies presented by the chemolithotrophs and the tendency for a given metabolic strategy to be distributed across several phylogenetic groups presents another challenge to the determination of chemolithotroph diversity in soils or sediments. For example, the majority of the sulphate-reducing bacteria form a relatively coherent phylogenetic group based on 16S rRNA sequences (Olsen *et al.*, 1994). However, recent information suggests that the sulphate-reducing bacteria are metabolically versatile and capable of using NO_3^- and Fe^{3+} reduction as well as sulphate reduction (Krekeler and Cypionka, 1995; Lovley, 1995). In addition, some sulphate-reducing bacteria may disproportionate S^o (forming both sulphate and sulphide; Lovley and Phillips, 1994) - a reaction that may be enhanced if a chemical sink for the sulphide formed (for example, FeOOH) is available. Therefore, an assessment of the diversity of nitrate-reducing chemolithotrophs might involve plans for the enrichment and isolation of the groups listed in Table 2.1, plus a variety of sulphate-reducers and numerous chemolitho-heterotrophic nitrate-reducers not listed in Table 2.1.

Suggestions to Overcome Technical Limitations

It is unlikely that the limitations of the enrichment culture method can be overcome - they are inherent to the process. However, it is important to realize that most discoveries in microbiology have depended on enrichment culture and the persistence and creativity of the investigator. One of the major impediments to our understanding of bacterial diversity is the limited number of individuals trained to isolate and describe bacteria. Expertise in particular bacterial groups typically resides with only a few individuals (in some cases, a single individual) worldwide. It is reasonable to assume that microbiologists with adequate training and basic tools, such as a microscope and the facilities to prepare culture media, can isolate new microorganisms belonging to the groups listed in Table 2.1. This is especially so if creative enrichment techniques, designed to simulate the *in situ* habitat, are employed. It is also possible that new genera or even families of bacteria might be isolated, and new types of chemolithotrophic metabolism identified. Establishment of a network of trained individuals is an important component of any protocol for determining bacterial diversity in soils and sediments.

Improvements in diversity assessment for chemolithotrophic bacteria could also be made by actively exploring unique habitats and by re-assessing mundane ones. Unique habitats might include environments with unique mineralogies, extreme physical or chemical conditions, unique juxtapositions of environmental gradients, or any of a variety of habitats restricted to specific global locations. Investigators should be open to the possibility of alternative chemolithotrophic metabolic schemes, based on the unique properties of specific habitats. Symbiotic associations with macrofauna should be considered, based on recent discoveries in marine systems. Finally, chemolithotrophic metabolism might be found in bacteria isolated under different metabolic regimes and from common habitats. The discovery of chemolithotrophy in diverse genera better known for alternative types of metabolism assists in drawing evolutionary relationships between those groups, especially when phylogenetic information from DNA- or RNA-based methods, or chemotaxonomic approaches, is also available.

New molecular and chemotaxonomic approaches which have been developed in recent years have provided new insights into the habitats and diversity of chemolithotrophic bacteria. The identification and determination of patterns of association of sulphur-oxidizing chemolithotrophs with marine invertebrates, determined solely by DNA- and RNA-based approaches (i.e. none of these organisms has been isolated in pure culture), are a good example (Distel *et al.*, 1994). Immunological approaches are also valuable. A monoclonal-based enzyme-linked immunosorbent assay (ELISA) was recently used to detect *Thiothrix* spp. in a variety of aquatic samples (Brigmon *et al.*, 1995) and similar procedures have been important in studies of ammonia-oxidizing bacteria. Chemotaxonomic procedures based on analysis of membrane lipids and their fatty acid methyl esters are also important (Ratledge and Wilkinson, 1988). All these methods require sophisticated equipment, are relatively costly and may suffer from bias as significant as those of enrichment culture (Moyer *et al.*, 1995; Tiedje and Zhou, this volume Chapter 5). Nevertheless, it is likely that comprehensive biodiversity assessments for any prokaryotic organisms must include many, if not all, of these approaches (O'Donnell *et al.*, 1994; Tiedje, 1994; Tiedje and Zhou, this volume Chapter 5).

Integrating Techniques into a General Scheme of Analysis

Several schemes for the assessment of microbial diversity in soils or sediments have been previously published (Freckman, 1994; O'Donnell *et al.*, 1994; Tiedje, 1994). An ideal soil or sediment biodiversity assessment would be a co-ordinated effort of biologists and soil or aquatic scientists. The field microbiologist would be part of a team of scientists investigating the diversity of a particular habitat, and all the physical, chemical and biological information about the habitat would be recorded and available for interpretation by interested parties. Database establishment and operation would be an integral part of the assessment activities (see Tiedje and Zhou, this volume Chapter 5). Microbiologists in the field would be trained to isolate bacteria and to be attuned to their ecology and physiology, so that new organisms, new habitats and new processes were discovered. Ideally, new isolates could be sent to large culture collections, or centres, where they would be characterized formally using technical procedures difficult to carry out in field locations. The ecological and physiological information obtained by the field microbiologist would accompany the isolate, or would be available as needed, so that new patterns of distribution and abundance could be discerned and ecological questions addressed in more detail using DNA- or RNA-based, immunological, chemotaxonomic or other approaches.

Bibliography

Ahmann, D., Roberts, A.L., Krumholz, L.R. and Morel, F.M.M. (1994) Microbe grows by reducing arsenic. *Nature* 371, 750.

Atlas, R.M. (1993) *Handbook of Microbiological Media*. CRC Press, Boca Raton.

Balashova, V.V. and Zavarzin, G.A. (1980) Anaerobic reduction of ferric iron by hydrogen bacteria. *Microbiology* 48, 635-639.

Balows, A., Trüper, H.G., Dworkin, M., Harder, W. and Schleifer, K.H. (eds) (1992) *The Prokaryotes. A Handbook on the Biology of Bacteria, Ecophysiology, Isolation, Identification, Applications,* 2nd edn. Springer-Verlag, New York.

Both, G.J., Gerards, S. and Laanbroek, H.J. (1992) The occurrence of chemolithoautotrophic nitrifiers in water-saturated grassland soils. *Microbial Ecology* 23, 15-26.

Brigmon, R.L., Bitton, G., Zam, S.G. and O'Brien, B. (1995) Development and application of a monoclonal antibody against *Thiothrix* spp. *Applied and Environmental Microbiology* 61, 13-20.

Brock, T.D. and Madigan, M.T. (1988) *Biology of Microorganisms,* 5th edn. Prentice Hall, New York.

Caccavo, F.Jr., Blakemore, R.P and Lovley, D.R. (1992) A hydrogen-oxidizing, Fe(III)-reducing microorganism from the Great Bay Estuary, New Hampshire. *Applied and Environmental Microbiology* 58, 3211-3216.

Caccavo, F.Jr., Lonergan, D.J., Lovley, D.R., Davis, M., Stolz, J.F. and McInerney, M.J. (1994) *Geobacter sulphurreducens* sp. nov., a hydrogen- and acetate-oxidizing dissimilatory metal-reducing microorganism. *Applied and Environmental Microbiology* 60, 3752-3759.

Cavanaugh, C. (1994) Microbial symbiosis, patterns of diversity in the marine environment. *American Zoologist* 34, 79-89.

Chapelle, F.H. (1993) *Ground-Water Microbiology and Geochemistry.* John Wiley and Sons Inc., New York.

Distel, D.L., Felbeck, H. and Cavanaugh, C.M. (1994) Evidence for phylogenetic congruence among sulphur-oxidizing chemoautotrophic bacterial endosymbionts and their bivalve hosts. *Journal of Molecular Evolution* 38, 533-542.

Emerson, D. and Revsbech, N.P. (1994) Investigations of an iron-oxidizing microbial mat community located near Aarhus, Denmark: field studies. *Applied and Environmental Microbiology* 60, 4022-4031.

Fisher, C.R. (1990) Chemoautotrophic and methanotrophic symbioses in marine invertebrates. *Reviews in Aquatic Sciences* 2, 399-436.

Freckman, D.W. (1994) *Life in the Soil. Soil Biodiversity: Its Importance to Ecosystem Process*. Report of a workshop held August 30-September 1, 1994 at the Natural History Museum, London, England. Colorado State University, Ft. Collins, Colorado.

Gerhardt, P., Murrary, R.G.E., Wood, W.A. and N.R. Krieg (eds) (1994) *Methods for General and Molecular Bacteriology*. American Society for Microbiology, Washington DC.

Ghiorse, W.C. and Ehrlich, H.L. (1993) Microbial biomineralization of iron and manganese. In: Fitzpatrick, R.W. and Skinner, C.W. (eds) *Iron and Manganese Biomineralization Processes in Modern and Ancient Environments*. Catena, Cremlingen-Destedt, pp. 75-99.

Kelly, D.P. (1992) The chemolithotrophic bacteria. In: Balows, A., Trüper, H.G., Dworkin, M., Harder, W. and Schleifer, K.H. (eds.) *The Prokaryotes. A Handbook on the Biology of Bacteria: Ecophysiology, Isolation, Identification, Applications*, 2nd edn. Springer-Verlag, New York, pp. 331-343.

Krekeler, D. and Cypionka, H. (1995) The preferred electron acceptor of *Desulfovibrio desulfuricans* CSN. *FEMS Microbiology Ecology* 17, 271-278.

Laverman, A.M., Blum, J.S., Schaefer, J.K., Phillips, E.J.P., Lovley, D.R. and Oremland, R.S. (1995) Growth of strain SES-3 with arsenate and other diverse electron acceptors. *Applied and Environmental Microbiology* 61, 3556-3561.

Lovley, D.R. (1991) Dissimilatory Fe(III) and Mn(IV) reduction. *Microbiological Reviews* 55, 259-287.

Lovley, D.R. (1995) Microbial reduction of iron, manganese, and other metals. *Advances in Agronomy* 54, 175-229.

Lovley, D.R. and Phillips, E.J.P. (1994) Novel processes for anaerobic sulfate production from elemental sulfur by sulfate-reducing bacteria. *Applied and Environmental Microbiology* 60, 2394-2399.

Meincke, M., Krieg, E. and Bock, E. (1989) *Nitrosovibrio* spp., the dominant ammonia-oxidizing bacteria in building stones. *Applied and Environmental Microbiology* 55, 2108-2110.

Moyer, C.G., Dobbs, F.C. and Karl, D.M. (1995) Phylogenetic diversity of the bacterial community from a microbial mat at an active, hydrothermal vent system, Loihi seamount, Hawaii. *Applied and Environmental Microbiology* 61, 1555-1562.

Nealson, K.H. and Myers, C.R. (1992) Microbial reduction of manganese and iron: new approaches to carbon cycling. *Applied and Environmental Microbiology* 58, 439-443.

O'Donnell, A.G., Goodfellow, M. and Hawksworth, D.L. (1994) Theoretical and practical aspects of the quantification of biodiversity among microorganisms. *Philosophical Transactions of the Royal Society of London B* 345, 65-73.

Olsen, G.J., Woese, C.R. and Overbeek, R. (1994) The winds of (evolutionary) change: breathing new life into microbiology. *Journal of Bacteriology* 176, 1-6.

Paul, E.A. and Clark, F.E. (1989) *Soil Microbiology and Biochemistry.* Academic Press Inc., New York.

Peters, V. and Conrad, R. (1995) Methanogenic and other strictly anaerobic bacteria in desert soil and other oxic soils. *Applied and Environmental Microbiology* 61, 1673-1676.

Polz, M.F., Distel, D.L. Zarda, B., Amann, R., Felbeck, H., Ott, J.A. and Cavanaugh, C.M. (1994) Phylogenetic analysis of a highly specific association between ectosymbiotic, sulphur-oxidizing bacteria and a marine nematode. *Applied and Environmental Microbiology* 60, 4461-4467.

Ratledge, C. and Wilkinson, S.G. (eds) (1988) *Microbial Lipids, Vol. 1.* Academic Press, New York.

Schlegel, H.G. and Bowien, B. (eds) (1989) *Autotrophic Bacteria.* Science Tech Publishers, Madison.

Staley, J.T., Bryant, M.P. Pfennig, N. and Holt, J.G. (eds) (1989) *Bergey's Manual of Systematic Bacteriology.* Williams and Wilkins, Baltimore.

Steenbergen, C.L.M., Sweerts, J.-P.R.A. and Cappenberg, T.E. (1993) Microbial biogeochemical activities in lakes, stratification and eutrophication. In: Ford, T.E. (ed.) *Aquatic Microbiology.* Blackwell Scientific Publications, London, pp. 69-100.

Tiedje, J.M. (1994) Approaches to the comprehensive evaluation of procaryote diversity of a habitat. In: Allsopp, D., Colwell, R.R. and Hawksworth, D.L. (eds) *Microbial Diversity and Ecosystem Function.* CAB INTERNATIONAL, Wallingford, pp. 73-87.

Ward, D.M., Weller, R. and Bateson, M.M. (1990) 16S rRNA sequences reveal numerous uncultured microorganisms in a natural community. *Nature* 345, 63-65.

Wimpenny, J. (1993) Spatial gradients in microbial ecosystems. In: Guerrero, R. and Pedros-Alio, C. (eds) *Trends in Microbial Ecology.* Spanish Society for Microbiology, Barcelona, Spain.

Woese, C.R., Kandler, O. and Wheelis, M.L. (1990) Towards a natural system of organisms, a proposal for the domains *Archaea, Bacteria*, and *Eucarya. Proceedings of the National Academy of Sciences, USA* 87, 4576-4579.

Actinomycetes 3

ELIZABETH M.H. WELLINGTON AND IAN K. TOTH

Department of Biological Sciences, University of Warwick, Coventry CV4 7AL, UK.

Studying the Ecology of Actinomycetes in the Soil and Rhizosphere

For the detection of streptomycetes and other actinomycetes in soil, most studies rely on the use of the dilution plate technique. This provides little information about the distribution and form (spores or mycelia) or about growth in soil, the majority of colonies probably originating from detached spores or other resting propagules. Detection of hyphal growth may therefore be difficult, but some methods for enumeration can be used which rely on differential resistance of spores and mycelia to heat (*Micromonospora*), physical disruption (streptomycetes) and lysis prior to DNA extraction. However, the importance of direct examination of the soil should be emphasized particularly in combination with cell concentration methods.

Direct Examination Techniques

Immunofluorescence

Antibodies to a specific antigen on the wall of a bacterial cell, once conjugated to a fluorescent dye, become a powerful tool in the enumeration and identification of specific bacterial populations. Unlike direct plating, which counts only culturable cells and often underestimates numbers, immunofluorescence (IF) does not differentiate between culturable and non-culturable cells, and so may allow a more accurate enumeration of cells from a particular environment. IF can also be used to give information on whether cells exist as single cells or in microcolonies and, in the case of soil, whether cells are associated with particular particles or microhabitats. In addition to these factors size comparisons of cells can also be made. IF can be used to enumerate cells directly in soil but, because of the magnification needed, the area of the field of view is very small. Therefore, the amount of soil which can be examined is also very small. The antibodies can be raised to cells at any stage in their life cycle. Methods for detection of actinomycetes have involved both monoclonal antibodies (MAb) derived from mice (Wipat *et al.,* 1992) and polyclonal antibodies from rabbits (Ridell and Williams, 1983).

Preparation of antigens

Antigens may be present on both spores and mycelia (Wipat *et al.,* 1992), but it is advisable to use the morphological form to be detected in the soil as the original source of antigenic material. Biomass for inoculation can be prepared by harvesting using filtration or centrifugation followed by washing. The following method is one recommended for raising antibodies to spores.

- Wash the spore pellet in 0.01% (w/v) phosphate buffer (PBS), centrifuge at 1240 ***g*** and 4°C for 10 min, and resuspend the pellet in 40% (w/v) sucrose.
- Layer the spore suspension onto a 40%:50%:60% (w/v) sucrose step gradient and centrifuge at 82,700 ***g*** for 1 h at 20°C.
- Harvest the spore layer from the 40%:50% interface.
- Add the spores to 20 ml sterile distilled water and centrifuge at 1240 ***g*** for 10 min at 4°C, repeating twice, to wash the spores before resuspending the spores in PBS (pH 7.0).
- Determine spore protein concentration, for example using the Coomassie blue protein assay of Bradford, using bovine serum albumin as the protein standard.
- Calculate total protein in spore suspension and freeze at -20°C until required.

The choice of monoclonal or polyclonal antibodies depends on the specificity required and proposed use of the antibody.

Fluorescent antibody preparation (methods based on Postma *et al.,* 1988)

- To antiserum at 2°C, add crystalline fluorescein isothiocyanate (FITC) over a 1 h period and stir for 12 h continuously. Conjugates are prepared with the minimal dye-protein ratio necessary to produce optimal staining homologous antigens (usually between 1:40 and 1:100 dye to protein ratio).

- Pass the conjugated sample through a 'Sephadex G-25' column (Pharmacia Ltd) equilibrated with 0.01 M, pH 7.1 PBS, at an hourly flow rate of 4.5 ml cm^{-1}.
- Store the conjugate at -20°C.

N.B. Before large-scale use of the conjugated antiserum, various dilutions are used to stain the filter and an optimum dilution for staining obtained.

Fluorescent antibody staining

- To 10 g soil add 195 ml of dispersing solutions (demineralized water, 0.1% partially hydrolysed gelatin diluted in 0.1 M ammonium phosphate, or 0.1% sodium pyrophosphate).
- Blend twice in a 'Braun' (or similar) blender for 20 s at maximum speed with a 5 s interval before adding a flocculation agent (1 g $CaCl_2{:}2H_2O$ or 0.7 g $Ca(OH)_2/MgCO_3$) and then shake vigorously by hand for 2 min.
- Allow to settle for 30-60 min. For different soil types, a different combination of dispersing solutions and flocculation agents, at different volumes and concentrations, may be necessary for efficient separation of bacteria from the soil.
- Filter the supernatant through a polycarbonate membrane filter (0.4 μm pore size, 25 mm diam., Milipore Corp.) stained with Indian ink. (Soak the filter in pre-filtered Indian ink and rinse off in running water.)
- After washing with 10-20 ml sterile saline (0.85% (w/v) NaCl), apply 0.05-0.1 ml gelatine-rhodamine isothiocyanate solution (diluted in 1 ml saline) to the filter to prevent non-specific binding.
- Place the filter at 50-60°C until almost dry.

Preparation of gelatin-rhodamine isothiocyanate solution

- Sterilize an aqueous solution (2% w/v) of gelatin, adjusted to pH 10-11 with 1 M NaOH, at 121°C for 10 min. Re-adjust the autoclaved solution to the same pH.
- Conjugate the gelatin using the following procedure.
  - Dissolve the rhodamine isothiocyanate (RhITC) in a minimum amount of acetone to provide 8 g dye mg^{-1} gelatin.
  - Add this to the gelatin solution, and stir gently overnight (Bohool and Schmidt, 1968).
  - Stain the filter with optimal antiserum dilution and keep under humid conditions in the dark for 30 min.
  - Remove excess antiserum by passing 10-100 ml saline through the filter.
  - Enumerate the stained cells using an epifluorescence microscope.

Enumeration

Count fifty, thirty or twenty microscopic fields depending on the number of cells present in each field, i.e. 0-1, 1-5 or 5 cells, respectively. Then, the number of cells per gram (N) is calculated from the formula of Schmidt *et al.* (1974).

$$N = \frac{(Nf.A.D)}{aV}$$

where: Nf is the average number /microscopic field
A is the effective filtering area (cm^2)
D is the dilution factor
a is the area of the microscopic field (cm^2)
V is the volume in ml of supernatant passed through the filter.

Limitations

Several problems and limitations are associated with IF techniques including the following (Bohool and Schmidt, 1980):

1. antibody specificity,
2. autofluorescence and/or non-specific binding,
3. antigen stability under growth conditions and environments,
4. inability to differentiate between live and dead cells and
5. efficiency of recovery of desired cells from the sample.

In order to test the specificity of an antibody it is important to test a large number of diverse bacterial species and genera before, for example, growth of a particular strain in a non-sterile soil system can be monitored. Using the above membrane technique to enumerate a population of bacteria in soil, the problem of autofluorescence is greatly diminished compared with direct enumeration in soil. On the other hand, non-specific staining has been a problem, although use of the method of Bohool and Schmidt (1968; see above) has much reduced this problem. Finally, although loss of bacterial cells during flocculation may be a problem, this loss can be reduced by using a suitable combination of dispersing solutions and flocculating agents.

Oligofluors

Oligonucleotide probes are used for *in-situ* detection and identification of individual cells. These probes are targeted towards ribosomal RNA (rRNA) and can be labelled with either radioisotopes or fluorescent dyes. Such a probe/fluorescent dye complex is referred to as an oligofluor. In nature, bacterial cells are often metabolically inactive, and, since rRNA is correlated with activity of cells, probing natural populations *in-situ* does not result in significant hybridization signals. Therefore, this effect allows the detection of metabolically active bacteria, i.e. thosc found in environments rich in nutrients. This approach therefore offers an alternative to serological approaches in the analysis of natural bacterial communities (Hahn *et al.*, 1992).

Oligonucleotide probes

Ribosomal RNA-targeted oligonucleotide probes are synthesized with a primary amino group at the 5' end (Aminolink 2; Applied Biosystems, Foster City). The fluorescent dye tetramethylrhodamine isothiocyanate (Tritc; Research Organics, Cleveland), is covalently linked to the amino group, and the dye-oligonucleotide conjugate (1:1) is purified from unreacted components and stored at -20°C in doubled distilled water at a concentration of 50 ng μl^{-1} (Amann *et al.*, 1990). The DNA-specific dye 4',6-diamidino-2-phenylindole (DAPI) (Sigma Ltd.) is stored in solution (1 mg ml^{-1}) at -20°C.

Cell fixation

Fix cells of pure cultures in 4% (v/v) paraformaldehyde/PBS, pH 7.2-7.4 (fixation buffer), for 3-16 h, wash once in PBS and store in 50% (v/v) ethanol/PBS at -20°C until further use (Amann *et al.*, 1990).

In situ hybridization

For each sample, use the following procedure.

- Apply 1 µl and tenfold dilutions of either cell suspensions or soil isolates to gelatin coated slides (0.1% (w/v) gelatin, 0.01% (w/v) $KCr(SO_4)_2$) and allow to air-dry.
- Following dehydration in 50, 80 and 100% (v/v) ethanol for 3 min, hybridize each of the preparations in 50 µl hybridization buffer (0.9 M NaCl, 0.1% SDS, 20 mM Tris, pH 7.2) and 1 µl probe (50 ng) at 45°C for 1 h.
- After hybridization, wash the slides in hybridization buffer for 20 min at 48°C, rinse with distilled water and air-dry.
- Add 1 µl of a DAPI solution (1 µl ml^{-1}) to each sample, cover with 5 µl hybridization buffer, incubate for 5 min at room temperature, rinse with distilled water and air-dry.
- Examine preparations with a microscope fitted for epifluorescence (with filter sets for 400 and 580 nm).

Cell extraction from soil

- Extract cells from 1 g soil before cell fixation with either 10 ml 0.1% (w/v) sodium pyrophosphate, pH 7.2, or 10 ml 0.1% pyrophosphate and a drop of 'Nonidet P-40' (Sigma Ltd). Alternatively fix cells directly in the soil sample with 1-5 ml fixation buffer (4% (v/v) paraformaldehyde/PBS, pH 7.2-7.4) for 3-16 h).
- Mix samples on a vortex mixer for 10 s and keep on ice for 2 min to allow a separation of heavy soil particles from the supernatant.
- Remove the supernatants and re-extract the soil pellets with either 2 ml PBS or 2 ml fixation buffer.
- Centrifuge the combined supernatants of the pyrophosphate extractions at 500 ***g*** for 15 min, discard the supernatant and resuspend the pellet in 1 ml fixation buffer.
- After fixing for 3-16 h, centrifuge all samples at 8000 ***g*** for 5 min, wash in 1 ml PBS and resuspend the pellet in 1 ml 50% (v/v) ethanol/PBS and store at -20°C.
- Use 1 µl of each sample as well as tenfold dilutions for microscopic determination of cell recoveries after *in situ* hybridization.

Before the addition of hyphae-forming Gram-positive organisms to soil microcosms, e.g. species of *Streptomyces, Frankia*, etc., mildly sonicate to homogenize cell clumps and pretreat with 0.1% (w/v) lysozyme in PBS at room temperature for 15 min.

Natural populations in soil

- Incubate 1 g soil at 30°C for 16 h, after the addition of 0.2 ml LB medium.
- Fix and hybridize this sample, together with an unamended sample.

Amendment and incubation may allow the growth of specific organisms, therefore allowing detection.

Limitations

Binding of oligonucleotide probes to soil does not occur, eliminating the need for counter-staining and blocking techniques as used with fluorescent antibody techniques. Autofluorescence of soil organic compounds can, however, disturb detection of bacteria. Direct fixation followed by extraction with fixation buffer results in lower contamination than pyrophosphate extraction followed by fixation, the former keeping organic material in larger particles (and also minimizing loss of bacteria attached to soil particles during the extraction procedure). Detection limits may be as low as 10^2-10^3 cells g^{-1} soil depending on soil type and the amount of liquid needed to obtain soil suspension. However, hybridization signals of spores are not observed.

Pure cultures of *Streptomyces* show much lower permeability for oligonucleotide probes than cells grown in soil. It is important therefore to pretreat pure cultures of cells with lysozyme (as outlined above) but is not necessary to treat soil-grown streptomycetes in this way. Amendment and incubation of soils with LB, starch or chitin-based media allows the detection of previously metabolically inactive cells.

Scanning Electron Microscopy

Scanning electron microscopy is subject to the same limitations as light microscopy as many fields of view need to be examined to sample a representative range of soil particles. However, studies using the scanning electron microscope (SEM) have made a valuable contribution to studying the form and arrangement of actinomycetes in soil (Mayfield *et al.*, 1972; Wellington *et al.*, 1990). SEM gives a greater resolution than light microscopy, but precludes the use of stains and fluors. Therefore, it is suitablc for studies of groups with a characteristic morphology which aids recognition. In addition, the extent of mycelial colonization of given habitats can be determined only by this technique.

Glutaraldehyde fixation

- Take approx. 5 ml culture and dilute with 25% (v/v) glutaraldehyde to a final glutaraldehyde concentration of 2.5%. Leave in the tube overnight at room temperature.
- Centrifuge the cells at 1240 *g*, and resuspend the pellet in 5 ml distilled water. Wash three to four times. Resuspend the final pellet in 0.5 ml distilled water.
- Add a drop of the cell suspension to a cover slip, previously attached to a 2.5 cm aluminium pin stub (Agar Scientific Ltd) with 'Electrodag' (Acheson Colloids Co.) and allow to air-dry.
- Sputter coat for 20 s.

Acetone

- To an agar plate freshly inoculated with actinomycete, add a round cover slip at a 45° angle.
- After growth of the cells for two to three days onto the cover slip, remove it and attach to an SEM stub with 'Electrodag'.
- Place the whole stub into 10, 25, 50, 75, 90 and 100% (v/v) acetone for 30 min each.
- Allow to air-dry before sputter coating.

Quantification

- Place a stub in the SEM, locate the approximate position of the centre of the stub and note the SEM micrometer position.
- Adjust the position of the field of view to lie at one corner of a 1 cm^2 region centred on the stub (stubs approx. 2-5 cm diam).
- Set the magnification so that the field of view is 100 m^2. Alter the field of view in 1 mm steps both horizontally and vertically so that 100 views of the 1 cm^2 region of the stub is observed.
- Study each field of view carefully for signs of mycelial growth and differentiation.

Each field of view is classified as no soil, soil only, low growth, medium growth and high growth. The percentage number of observations are then noted (Clewlow *et al.*, 1990).

Extraction/Sampling Methods

Extraction of Spores and Mycelia

Most isolation techniques involve a cursory extraction method coupled with a dilution of biomass to allow culturing. In many cases, actinomycete genera are present in soil at $< 10^4$ colony-forming units (cfu) g^{-1}. Therefore, there is a need to develop methods which concentrate cells in a given sample and improve chances of detection. Soil biomass extraction can be either physical or chemical or a combination of both. Mycelia will become intimately associated with soil particles, and so there is an even greater need to carry out some form of dispersal to extract cells.

Physical disruption methods

Ringers extraction (Wellington *et al.*, 1990)

- Shake 1 g soil in 9 ml 1/4 strength Ringers solution on a rotary shaker for 10 min.
- Serially dilute supernatant and count on selective agar.

Homogenization

- Add 1 g soil to a 150 ml conical flask containing 99 ml sterile distilled water with 'Teepol' (a surfactant) at 10 mg l^{-1}.
- Attach the flask to a high-efficiency homogenizer drive, and surround the flask with crushed ice.

- Homogenize at high speed for 10, 20, 30, 60 and 90 s, to establish the period required to release maximum numbers of actinomycetes from a given soil.
- Sample at each time point, serially dilute and replica plate onto appropriate agar (Baecker and Ryan, 1987).

Chemical disruption method

Ion-exchange resin method (Herron and Wellington, 1990)

- Transfer 100 g soil to a 500 ml centrifuge pot, add 20 g of 'Chelex 100', 100 ml of 2.5% (v/v) polyethylene glycol (PEG 6000; Sigma Ltd) plus 0.1% sodium deoxycholate (w/v).
- Shake the pot gently with occasional fast agitation for 2 h at 4°C on a rotary shaker.
- Centrifuge for 30 s at 850 ***g*** and retain pellet.
- Filter the supernatant through a membrane filter holder (Millipore Corp.) with no filter present, the metal support being sufficient to remove any 'Chelex 100' from the liquid.
- Add a further 100 ml PEG/deoxycholate to the pellet and re-extract as before.
- After the second filtration, pool the supernatants and centrifuge at 2200 ***g*** for 15 min.
- Allow the resulting pellet to resuspend overnight in 1/4 strength Ringers solution before measuring the volume of the suspension and determining the viable count. Cell concentration techniques (see later) may also be used with extraction.

Extraction of actinophage (Lanning and Williams, 1982)

- To 20 g soil, add 50 ml nutrient broth (pH 8.0) containing 0.1% (w/v) egg albumin (Sigma Ltd) and shake for 30 min at 200 oscillations min^{-1} on a rotary flask shaker, and leave for 16 h at 4°C.
- Centrifuge the supernatant at 1240 ***g*** for 30 min and pass the resultant supernatant through a 0.45 μm pore size nitrocellulose filter.
- Count phage by overlaying 0.1 ml of lysate with soft nutrient agar seeded with the sensitive bacterial strain.

Limitations

The Chelex technique has been modified to allow extraction of 10 g soil by reducing all weights and volumes by a factor of 10. This allows Chelex extraction to be carried out on fed-batch microcosms without disruption to the system. The Chelex technique shows a significantly greater extraction efficiency for removal of *Streptomyces lividans* TK24 spores from soil than the use of Ringers (125.7% vs 16.2%) and inoculum of 2.5% PEG 6000 in the eluent yields significantly greater numbers of spores than does sodium deoxycholate alone (125.7% *vs.* 34.8%). As few as 10 spores can be detected when 100 g soil are inoculated with *S. lividans* TK24, but cannot be enumerated accurately because of large variations inherent in counting low numbers of colonies. Recoveries of both total cells (Ringers) and spores (Chelex) are less efficient from non-sterile than sterile soils.

Extraction of DNA from soil (see Cresswell *et al.,* 1992, and this volume, Chapter 5)

SDS/heat lysis

- Suspend 10 g soil in 20 ml sodium phosphate buffer (0.12 M, pH 8.0: SPB), containing 1.5 % (w/v) SDS.
- Heat samples to 70°C for 1 h, mixing occasionally and shake on a rotary flask shaker (maximum setting) for 10 min.
- Centrifuge at 1660 ***g*** for 10 min at 20°C and re-extract the pellet formed with 10 ml SPB.
- Pool the supernatants and centrifuge at 13,800 ***g*** for 20 min at 20°C.
- Recover the supernatant and precipitate the DNA with PEG 6000, then centrifuge at 2260 ***g*** for 10 min at 20°C.
- Resuspend the pellet in 5 ml TE-buffer and extract twice with equal volumes of TE-saturated phenol (Hopwood *et al.,* 1985).
- Separate the phases by centrifugation at 2260 ***g*** for 10 min at 20°C.
- Re-extract the phenol phase with TE-buffer, pool the aqueous phases and extract with chloroform/iso-amyl alcohol (24:1).
- Retain the upper aqueous phase and precipitate with 2.5 volumes ethanol overnight at -20°C.
- Centrifuge at 2260 ***g*** for 10 min at 20°C, and wash with ice-cold 70% (v/v) ethanol.
- Vacuum dry and dissolve in 100 ml TE-buffer by heating to 60°C, if necessary. Store at 4°C.

Bead beating

- Wash 10 g soil into a bead beating bottle (50 ml; B. Braun, Melsungen) using SPB, add approx. 10 ml 0.1-0.11 mm glass beads (0.1-0.11 mm; B. Braun, Melsungen) and fill the bottle to its neck with SPB (approx. 40 ml total volume).
- Place the bottle inside a bead beater (B. Braun, Melsungen) and shake for five 1 min periods with CO_2 cooling.

- Transfer the contents to an 'Oakridge' tube and pellet large soil particles and glass beads by centrifuging at 4500 ***g*** for 15 min. Retain the supernatant.
- Re-extract the pellet by vortexing for 1 min with a further 20 ml SPB, pool the supernatants and add 0.2 volumes of 8 M potassium acetate and place immediately on ice for 15 min.
- Centrifuge the mixture at 15,200 ***g*** for 30 min and transfer the supernatant equally into two 'Oakridge' tubes.
- Add 10 ml 50% (w/v) PEG to each tube and make up to 0.5 M with 5 M NaCl. Precipitate the DNA overnight at 4°C.
- Centrifuge at 8820 ***g*** for 10 min and dissolve the DNA pellet in 500 µl TE.
- Transfer the soil DNA extract to a large 'Eppendorf' tube and extract twice with phenol solution, followed by precipitation with 1 volume isopropanol at room temperature for 5 min.
- After pelleting by centrifugation at max. speed for 10 min, resuspend the DNA in TE buffer and re-precipitate with 25 µl 100 mM spermine-HCL for 5 min at room temperature.
- Wash the DNA in 1 ml 70% (v/v) ethanol and resuspend in 50 µl TE buffer (Cresswell *et al.,* 1991).

Limitations

Both SDS/heat lysis and bead beating yield similar amounts of plasmid DNA from mycelial inocula, allowing detection of 2.16×10^4 cfu g^{-1}. In extracts inoculated with spores, no plasmid DNA is detected using SDS/lysis. Therefore, two extraction methods allow differentiation between mycelial- and spore-borne DNA. At high levels of mycelial inocula, both methods appear to show similar efficiencies of DNA extraction, whereas at low levels bead beating is slightly more efficient, with a detection limit of as few as 10 or less cfu g^{-1} soil compared with 39 cfu g^{-1} for SDS/heat lysis. Whereas SDS/heat lysis is unable to detect plasmid DNA from spores, even at high inocula, bead beating is able to detect as few as 4.12×10^3 cfu g^{-1} soil. The low detection limits of plasmid DNA from mycelia compared with spores may be due to an increased copy number of plasmid in growing mycelia, which may be significantly less in spores (Wellington *et al.,* 1992).

Extraction of rRNA from Soil

Extraction of bacterial cells

- Suspend 10 g soil in 40 ml 0.1 M $Na_4P_20_7$, shake for 15 min and centrifuge at 1000 ***g*** to remove fungi and inorganic substances.
- Extract the pellet once more with 20 ml TE buffer.
- Centrifuge the pooled supernatants at 15,000 ***g*** and wash the resultant pellet twice with TE-buffer to remove soluble amounts of humic acids.

This bacterial fraction, still contaminated with humic acids, is used for rRNA isolation.

Ribosomal RNA extraction

- Sonicate either pure bacterial cultures, or the extracted bacterial fractions of 10 g soil, or 2 g (wet weight) soil, three to six times for 30 s each at 60 W in 5 ml of 7.5 M guanidine-hydrochloride/1 M Tris (pH 7.0) and centrifuge at 3000 ***g*** to separate insoluble residues.
- Precipitate soluble substances of the supernatant with ethanol overnight at -20°C, resuspend the ethanol-insoluble pellet in sterile distilled water and extract twice with phenol/chloroform solution (pH 7.4).
- After subsequent chloroform extraction (× 2), precipitate the colourless rRNA with ethanol, dry and resuspend in distilled water.
- If the pellet is brown resuspend it in 0.4 ml guanidine-hydrochloride solution, precipitate with ethanol and resuspend the resulting pellet in distilled water and extract with phenol/chloroform. This procedure is repeated until the pellet is colourless.

Limitations

Ribosomal RNA isolation directly from soil yields about 10 ng rRNA g^{-1} soil (from cfu of 1.4×10^6 to 1.2×10^7) depending on soil type. The total yield of rRNA is usually much lower when bacteria are first separated from soil. This difference in yield is probably due to the amount of washing steps used to remove humic acids from bacterial suspensions.

Cell Concentration Techniques

Immunomagnetic capture

Immunomagnetic capture relies on the ability of bacterial cells to attach to beads coated in a specific antibody to those cells. The three main requirements for such a system are: first, to obtain an antibody to a specific cell surface component; second, to prevent non-specific attachment of cells to the beads; and, third, to obtain bead-cell complexes that are sufficiently stable to be attracted towards a magnet. Immunomagnetic capture is an important method as an initial step in the isolation of genetically modified organisms (GMOs), a method which may then be followed by other direct detection techniques to determine the presence of specific marker genes.

Immunomagnetic capture has been carried out on a number of bacterial genera including *Pseudomonas* (Morgan *et al.*, 1991), *Salmonella*, *Listeria* (Skjerve *et al.*, 1990) and *Streptosporangium* (Mullins *et al.*, 1995). This technique has also been used for separating *Streptomyces* from both buffer and soil samples, and is briefly outlined below. Wipat *et al.* (1992) used such a system to separate *Streptomyces lividans* spores from both phosphate-buffered saline and non-sterile soil extracts. Magnetized polystyrene beads ('Dynabeads') are coated in spore-specific antibody (43H6) and used to recover selectively spores of cluster 21 species. It has been shown that recovery values increase as the final bead concentration in a sample is raised.

Method based on Pseudomonas putida separation (Morgan *et al.*, 1991)

Preparation of MAb

- (Method as described previously.) Use the manufacturer's recommended procedures for coupling MAb to the 'Dynabead M-450' particles (Dynal, New Ferry, Wirral).

Agglutination test

- Apply 20 µl of bead suspension (108 beads ml^{-1}) to a slide with an equal volume of bacterial/spore suspension.
- After 60 s of incubation with gentle rocking, aggregates are clearly visible as dark grains in positive samples.
- To determine cell attachment perform phase-contrast microscopic examination on wet preparations in conjugation with fluorescent microscopy on samples stained for 5 min with 0.05% (w/v) acridine orange.

Cell separations

- Add 100 µl of X10 PBS to each test tube containing 1 ml bacterial sample, and gently mix.
- Add 30 µl of bead suspension to each sample and incubate for 15 min at 20°C.
- Attract beads and bead-cell complexes to the side of each test tube by placing the test tube in a magnetic particle concentrator (Dynal).
- Pipette off the solution and replace it with 1 ml PBS (1 ×) as a washing step. Repeat this two to three times.
- For scanning electron microscopy, resuspend bead samples in distilled water after the final wash.
- To determine cell attachment observe bead samples by phase-contrast microscopy (or fluorescence microscopy with acridine orange-stained samples). In order to perform cell culture, bead-cell complexes diluted in distilled water are either vortexed for 2 min or sonicated at an amplitude of 10 µm, peak to peak, for 10 s.

Limitations

To reduce non-specific attachment of cells to the beads, detergents may be included in all solutions during cell separation. Although this severely reduces the colony-forming ability of cells they remain intact and can be monitored by direct detection methods. To further reduce background contamination, an optimum ratio of glass beads to target cells would need to be added to the sample.

If the number of cells cannot be estimated, an excess of glass beads is often preferred. Another method of reducing the level of non-specific binding is an increase in the number of wash steps. However, this would rely on the need for strong antigen binding capable of tolerating shear forces due to washing. An increase in antigen binding may be achieved by attachment of the beads to several surface points.

The isolation of cells from solid medium, e.g. soil or silt, creates problems. First, antigen-bead interactions would need to be stronger than the cell-particle association before the cells could be removed. Second, if target cells were present within microcolonies on these particles, both access to the antigen and detachment of target cells from the microcolony would have to occur for a successful separation (Morgan *et al.,* 1991).

Model Soil Systems

The natural environment is a dynamic system of continually changing conditions. Soil is a complex structure, which has been reviewed elsewhere. It is a nutrient-poor environment which is exposed to large variations in factors such as pH, temperature, nutrient availability and atmospheric conditions. The solid constituency of soil leads to limited movement of water, gases, nutrients and organisms within it. This in turn leads to heterogeneity within the environment and therefore to the diversity of microorganisms within it. Due to the particulate nature of soil, gradients appear between the atmosphere and the soil. These gradients are caused by differences in relative rates of consumption and production of organic matter and gases, and the rates of physical/chemical interactions between organisms/inorganic solids and percolating water (Nedwell and Gray, 1987). Over time these gradients result in the formation of many different microenvironments for microbial growth. For example, members of the genus *Streptomyces* are aerobic and, although low numbers of streptomycetes are often found in waterlogged, anaerobic or acid soils, much greater levels are found in well aerated, neutral soils.

Even from this brief introduction, it is clear that the design and running of a laboratory-based soil microcosm system poses many problems. Important factors which must be regulated in order for such a system to succeed include water/air ratios, pH, temperature, nutrient and growth matrix availability. In order to study the effects of one or more of these variables on the survival of an actinomycete, microcosms must be set up under defined conditions. However, because of the complex interactions which occur at individual microsites, it would be an impossible task to eliminate other variables altogether.

By far the largest proportion of streptomycetes in soil exist as spores (> 90%). In this state the cells remain dormant. Only in the mycelial state do factors such as gene transfer, phage infection, the effects of antibiotics, etc. play a role in the life cycle of the cell. For this reason it is important to develop microcosms which allow continuous rounds of germination and sporulation. A number of laboratory microcosms have been designed which allow the growth of streptomycetes. These systems include the batch microcosm (Bleakley and Crawford, 1989; Wellington *et al.*, 1990) and the fed-batch microcosm (Cresswell *et al.*, 1992; Marsh *et al.*, 1993).

Soil collection and preparation

- Collect samples from the top 20 cm of soil, place for two to three weeks at room temperature to air dry, then grind the soil in a mortar and pestle and sieve through a 2 mm mesh.

Sterilization and amendment of soil

- Autoclave the soil in a sealed container at 121°C for 15 min, allow to stand for 24 h at room temperature and re-autoclave (this step is necessary for complete elimination of spores).
- Amend the sterile soil with sterile distilled water to a moisture content equivalent to 40% water-holding capacity (WHC).
- Nutrient amendments may also be made with 1% (w/w) soluble starch and 1% (w/w) chitin (from crab shell, Sigma Ltd), although autoclaving has been shown to release nutrients from the soil.

The amendment procedure may also be used with non-sterile soil at 40% WHC, if desired.

Inoculation of microcosms

Microcosms are inoculated with spores, with mycelia or with sterile soil previously inoculated with spores and incubated for three days.

Preparation of spores for inoculation

- Streak approximately ten agar plates with the streptomycete and incubate for three to four days until sporulation occurs.
- Remove the spores from the mycelia by adding 5 ml 1/4 strength Ringers solution to each plate, followed by gentle agitation with a glass spreader.

- Pour the crude suspensions into a suitable container and vortex for 1 min.
- Filter the suspension through non-adsorbent cotton wool using a sterile syringe.
- Pour the filtered suspension into a centrifuge tube and centrifuge for 5-10 min at 2200 ***g*** to pellet the spores.
- Immediately pour off the supernatant and agitate the container to disperse the pellet in the remaining liquid.
- Add sterile 20% (v/v) glycerol (1-2 ml) and briefly agitate again.
- Freeze the sample at -20°C until required (adapted from Hopwood *et al.*, 1985).

Preparation of mycelia for inoculation

- Add spores to 10 g of sterile soil and incubate at 21°C for three days.
- Add the incubated soil to the microcosm and mix with a sterile spatula.

Batch microcosms

- Add 10-100 g volumes soil (sterile or non-sterile) to a glass beaker allowing a soil depth of 2-5 cm.
- Amend and inoculate the microcosm as appropriate, and seal the microcosm with aluminium foil, before incubating at 21°C until sampling. Once sampled, discard the microcosms (Herron and Wellington, 1990).

Fed-batch microcosms

- Add 200 g soil (sterile or non-sterile) to a glass beaker, to produce a soil depth of 2-5 cm.
- Amend and inoculate the microcosm as outlined above, and seal the microcosm with aluminium foil before incubating at 21°C.
- On day fifteen, and every subsequent fifteenth day up to day sixty, mix the microcosms and replace 50% of the weight of the soil with fresh, uninoculated, nutrient-amended soil.
- Use the removed soil for extraction of biomass and nucleic acids. For samples intermediate of the above, microcosms are set up and destructively sampled without soil replacement (Cresswell *et al.*, 1992).

Improved fed-batch microcosm

- Add 500 g soil (sterile or non-sterile) to a sealable plastic box (containing small air holes covered with muslin), to produce a soil depth of 2-5 cm.
- Amend and inoculate the microcosm as outlined above, before incubating at 21°C.

- On day ten and every subsequent tenth day, remove 100 g of the soil and replace with amended, uninoculated soil.
- Use the removed soil for extraction of biomass and nucleic acids.
- On intermediate days remove samples from the microcosm using a sterilized cork borer (the holes created by the borer are left unfilled; Marsh *et al.*, 1993).

Limitations of Existing Methods

With the improved fed-batch microcosm Chelex extractions are carried out using only 10 g soil in order to prevent excess loss of soil from the microcosm. An important factor in the success of this system is the use of a sealable plastic box, which prevents water loss from the microcosm, and therefore the moisture content of the soil remains constant, reducing variability, within the system, over time.

Bibliography

Amann R.I., Binder, B.J., Olson, R.J., Chisholm, S.W., Devereux, R. and Stahl, D.A. (1990) Combination of 16S rRNA-targeted oligonucleotide probes with flow cytometry for analysing mixed microbial populations. *Applied and Environmental Microbiology* 56, 1919-1925.

Baecker, A.A.W. and Ryan, K.C. (1987) Improving the isolation of actinomycetes from soil by high-speed homogenization *South African Journal of Plant and Soil* 4, 165-170.

Bleakley, B.H. and Crawford, D.L. (1989) The effects of varying moisture and nutrient levels on the transfer of a conjugative plasmid between *Streptomyces* species in soil. *Canadian Journal of Microbiology* 35, 544-549.

Bohool, B.B. and Schmidt, E.L. (1968) Nonspecific staining: its control in immunofluorescence examination of soil. *Science* 162, 1012-1014.

Bohool, B.B. and Schmidt, E.L (1980) The immunofluorescence approach in microbial ecology. In: Alexander, M. (ed.) *Advances in Microbial Ecology*, vol. 4. Plenum Press, New York, pp. 203-241.

Clewlow, L.J., Cresswell, N. and Wellington, E.M.H. (1990) Mathematical model of plasmid transfer between strains of streptomycetes in soil. *Applied and Environmental Microbiology* 56, 3139-3145.

Cresswell, N., Saunders, V.A. and Wellington, E.M.H. (1991) Detection and quantification of *Streptomyces violaceolatus* plasmid DNA in soil. *Letters in Applied Microbiology* 13, 193-197.

Cresswell, N., Herron, P.R., Saunders, V.A. and Wellington, E.M.H. (1992) The fate of introduced streptomycetes, plasmid and phage populations in a dynamic soil system. *Journal of General Microbiology* 138, 659-666.

Hahn, D., Amann, R.I., Ludwig, W., Akkermans, A.D.L. and Schleifer, K.H. (1992) Detection of micro-organisms in soil after *in situ* hybridization with rRNA-targeted, fluorescently labelled oligonucleotides. *Journal of General Microbiology* 138, 879-887.

Herron, P.R. and Wellington, E.M.H. (1990) New method for extraction of streptomycete spores from soil and application to the study of lysogeny in serile amended and non-sterile soil. *Applied and Environmental Microbiology* 56, 1406-1412.

Hopwood, D.A., Bibb, M.J., Chater, K.F., Kieser, T., Bruton, C.J., Kieser, H.M., Lydiate, D.J., Smith, C.P., Ward, J.M. and Schrempf, H. (eds) (1985) *Genetic Manipulation of Streptomyces - A Laboratory Manual.* The John Innes Institute, Norwich, England.

Lanning, S. and Williams, S.T. (1982) Methods for the direct isolation and enumeration of actinophages in soil. *Journal of General Microbiology* 128, 2063-2071.

Marsh, P., Toth, I., Meijer, M., Schilhabel, M.B. and Wellington, E.M.H. (1993) Survival of the temperate actinophage yenC31 and *Streptomyces lividans* in soil and the effects of competition and selection on the spread of lysogens. *FEMS Microbiology Ecology* 13, 13-21.

Mayfield, C.I., Williams, S.T., Ruddick, S.M. and Hatfield, H.L. (1972) Studies on the ecology of actinomycetes in soil. IV. Observations on the form of growth of streptomycetes in soil. *Soil Biology and Biochemistry* 4, 79-91.

Morgan, J.A.W., Cranwell, P.A. and Pickup, R.W. (1991) Survival of *Aeromonas salmonicida* in lake water. *Applied and Environmental Microbiology* 57, 1777-1782.

Mullins, P.M., Gurtler, H.G. and Wellington, E.M.H. (1995) The selective recovery of *Streptosporangium fragile* from soil by indirect immunomagnetic capture. *Microbiology* 141, 2149-2156.

Nedwell, D.B. and Gray, T.R.G. (1987) Soils and sediments as matrices for microbial growth. In: Fletcher, M., Gray, T.R.G. and Jones, J.G. (eds) *Ecology of Microbial Communities*, No. 4, Cambridge University Press, Cambridge, pp. 21-54.

Postma, J., van Elsas, J.D., Govaert, J.M. and van Veen, J.A. (1988) The dynamics of *Rhizobium leguminosarum* biovar *trifolii* introduced into soil as determined by immunofluorescence and selective plating techniques. *FEMS Microbial Ecology* 53, 251-260.

Ridell, M. and Williams, S.T. (1983) Serotaxonomical analyses of some *Streptomyces* and related organisms. *Journal of General Microbiology* 129, 2857-2861.

Schmidt, E.L., Biesbrock, J.A., Bohlool, B.B. and Marx, D.H. (1974) Study of mycorrhizae by means of fluorescent antibody. *Canadian Journal of Microbiology* 20, 137-140.

Skjerve, E., Rcrvik, L.M. and Olsvik, O. (1990) Detection of *Listeria monocytogenes* in foods by immunogenetic separation. *Applied and Environmental Microbiology* 56, 3478-3481.

Wellington, E.M.H., Cresswell, N. and Saunders, V.A. (1990) Growth and survival of streptomycete inoculants and extent of plasmid transfer in sterile and non-sterile soil. *Applied and Environmental Microbiology* 56, 1413-1419.

Wellington, E.M.H., Cresswell, N. and Herron, P.R. (1992) Gene transfer between streptomycetes in soil. *Gene* 115, 193-198.

Wipat, A., Wellington, E.M.H. and Saunders, V.A. (1992) Detection systems for streptomycetes. In: Wellington, E.M.H. and van Elsas, J.D. (eds) *Genetic Interactions Between Microorganisms in the Natural Environment.* Pergamon Press, Oxford, pp. 83-90.

Cyanobacteria (Blue-Green Algae) 4

Brian A. Whitton

Department of Biological Sciences, University of Durham, Durham DH1 3LE, UK.

The Importance of Cyanobacteria in Ecosystem Processes

The cyanobacteria are by far the largest group of phototrophic prokaryotes, occurring in a wide range of habitats. Much of the Earth's original atmospheric oxygen was probably formed by organisms quite like modern cyanobacteria and they are still responsible for a considerable proportion of oceanic photosynthetic oxygen evolution. Their key features are given in most general algal texts, but the account by Castenholz and Waterbury (1989) is especially useful.

Their success appears to be due to a number of features widespread in the group (Whitton, 1992), of which the following are likely to be important in soils and sediments. The temperature optimum for many cyanobacteria is higher by at least several degrees than most eukaryotic algae (Castenholz and Waterbury, 1989). Many terrestrial forms tolerate high levels of ultraviolet irradiation. The success of many planktonic forms is favoured by their ability to use light efficiently at low light intensities and this may, perhaps, apply also to some soil forms. Tolerance of desiccation and water stress is widespread and cyanobacteria are among the most successful organisms in highly saline environments (Borowitzka, 1986). Free sulphide is tolerated by some cyanobacteria at levels much higher than those tolerated by most eukaryotic algae. The ability of many species to fix N_2 provides a competitive advantage where levels of combined nitrogen are low. There are many reports of N_2-fixing cyanobacterial growths on or in soils, including desert crusts, agricultural soils and rice-field ecosystems (Whitton, 1993).

Cyanobacteria frequently cover the surfaces of buildings and rocks, especially in the tropics, and may play an important role in damage to these surfaces or in the erosion of calcareous rocks. In other circumstances they may play a protective role in reducing erosion of soils and dunes.

Naming Cyanobacteria

At present, cyanobacteria can be named according to two different systems: the International Code of Botanical Nomenclature (ICBN) and the International Code of Nomenclature of Bacteria (ICNB). The work already underway to harmonize the codes of biological nomenclature will alleviate this problem, if successful. Although many cyanobacterial forms are well-known and easy to recognize, it is often difficult to identify all the material taken from a site in nature and, even more so, mixed material growing in culture. Many species are ill-defined and individual strains are highly variable, so frequently there is a subjective element in the application of binomials. Nevertheless, nearly all names introduced under the botanical system are validated using descriptions based on observable morphological features and are suitable for studies of the diversity of field samples, whereas names introduced under the bacteriological system are validated using a wide range of features present in isolates originating from single cells or filaments.

Unfortunately, there is no text focusing on the identification of species from soils and sediments and those general accounts which do exist are not particularly well suited to naming soil forms, and are often difficult to obtain. For many people, the account by Desikachary (1959) is still the best starting place for naming species. Although this deals specifically with taxa from India, many species appear to be cosmopolitan or distributed very widely. However, biodiversity studies on cyanobacteria will remain difficult until a modern account is produced.

Cyanobacteria should be named from live material wherever possible. It is often difficult, or even impossible, to name preserved material using available techniques and published data. Most soil forms can be dried and re-wetted easily, so there should seldom be any need to deal with preserved material.

Cyanobacteria in Soils and Sediments

Simple enrichment culture techniques show the presence of cyanobacteria in the great majority of soils at pH values of 6.5 or greater and, especially in the tropics, sometimes at much lower values. There are no records for any habitat below pH 4.0. Provided that they are supplied with a suitable inorganic medium, the majority of soil cyanobacteria grow quite easily on agar. Well-aerated marine communities, such as growths on rocks and coarse sediments, can be treated in a similar way to soil, except that seawater should be incorporated rather than distilled water (see below). More complex communities overlying sediments with reduced oxygen levels near the surface, especially well-developed intertidal mats, are much more difficult to study in enrichment culture, and it is important to study live natural material as well as cultures.

Even where a standard protocol is suited for studying biodiversity, as with most soils, it is important to be aware of several features of cyanobacterial biology.

Many soil and sediment cyanobacteria (especially the former) form discrete colonies when plated on agar, and there are several reports on the number of colony-forming units per unit area in rice-field soils. However, others are usually seen on agar plates only as part of mixed communities. A problem arises in bringing such organisms to a condition suitable for taxonomic identification. Many genera and species are characterized by morphological features which are expressed only when the organisms are stressed by one or more environmental features, especially nitrogen or phosphorus limitation. On the one hand, it is necessary to provide sufficient nutrients to permit the organisms to produce a large enough mass to permit recognition, while, on the other, they must be grown to a particular nutrient limitation. At the same time, they should not be overwhelmed by faster-growing 'weedy' species.

Standard Protocol for Soil

Background

Information on environmental conditions in nature (especially temperature, light flux at the soil surface and pH) is helpful is establishing optimum conditions for isolation. However, it seems likely that the wider the range of physical and inorganic chemical conditions tested, the greater the diversity of cyanobacteria will be shown. The following protocol describes the minimum procedure required to reveal most organisms, and then lists minor variations on this procedure which may reveal a few more species. The general procedure may be applied to any non-saline soil in the pH range 6.5-9.0. Modifications for low pH, saline and other soils, sediments and other types of environment are described later.

The activity of cyanobacterial pathogens is apparently relatively unimportant in soil isolation studies on agar, although cyanophages are important in some aquatic environments. Competition by eukaryotic algae and harmful effects of fungi may be reduced by adding cycloheximide, but grazing by soil protozoa (especially amoebae) unharmed by cycloheximide probably sometimes has a selective effect on the species found.

Safety precautions

Cycloheximide is very toxic if swallowed, and it is also important to avoid contact with skin, so gloves must be worn. Should skin contact be made, wash thoroughly with water.

Sampling the Material

- If there are visually obvious growths of phototrophs on the soil surface (e.g. cyanobacteria, algae or moss protonema), remove much of these. Retain a small sample to check the cyanobacteria and discard the remainder.
- Take a set of samples from the top 5 mm of soil. Ideally, at least 10 samples should be taken from different microenvironments, mixed and approx. 5 g taken for further study.
- After the sample has been removed from the field site, care should be taken to avoid contamination by organisms in the air.
- If the material is to be investigated further within the next 24 h, there is no need to do more. If the delay is longer, the soil should be air-dried before storage. In order to avoid the risk of further contamination, it may be necessary to do this in a laminar flow cabinet, but never use a desiccant such as silica gel.
- Samples may then be stored for extended periods before investigating biodiversity. The greater the extent to which the soils are ones subjected to pronounced drying in nature (especially alternating wet-dry periods), the slower is likely to be the rate of loss of species shown in biodiversity assays. (Alternating wet-dry, calcareous soils can show high diversity at least 25 years after sampling.)

Preparation of Media and Agar Plates

The protocol requires the use of three different types of inorganic medium made up with distilled water. The media have many components in common, but are modified to provide: **A** N-free medium, **B** low N:P ratio medium, **C** high N:P ratio medium. These three media contain (mg l^{-1}): **A** P, 1; **B** N, 4; P, 1; **C** N, 8; P, 0.2. Medium **A** is designed to favour N_2-fixers, so it contains twice the Fe content of media **B** and **C**. Because of the need to adjust the N and P contents, the three media also differ slightly from one another with respect to the balancing ions Na, K and Cl.

The media consist entirely of inorganic components, apart from the chelating agent (EDTA, ethylenediaminetetra-acetic acid) and buffer (HEPES, N-2-hydroxymethylpiperazine-N'-2-ethane sulphonic acid). About 2 l of each of the three media are required, but it is recommended to make a slight surplus.

Preparation of stock solutions

Prepare stock solutions of salts containing the major elements, together with stock solutions of Fe-EDTA and trace metals.

Fe-EDTA Prepare 100 ml stock solution containing: 97 g $FeCl_3.6H_2O$ + 1.335 g Na_2EDTA

Trace metals Prepare 100 ml stock solution containing:
0.286 g H_3BO_3 0.181 g $MnCl_2.4H_2O$ 0.022 g $ZnSO_4.7H_2O$
0.039 g $Na_2MoO_4.2H_2O$ 0.0079 g $CuSO_4.5H_2O$ 0.0049 g $Co(NO_3)2.6H_2O$
0.0048 g $NiSO_4.7H_2O$

Preparation of media

The formula given is that for one litre of medium. If the standard isolation protocol is followed (needing 48 Petri dishes of each medium), then twice the volume of each medium (i.e. 2 l) should be prepared.

The standard procedure involves making 500 ml buffer solution, adding about 400 ml distilled water, followed by the various stock solutions, adjusting the pH to the required value with NaOH and adjust the volume to 1 l. When adding the stocks, ensure that Fe-EDTA is added before trace metals.

For each litre of final medium, dissolve 0.6 g HEPES in 500 ml distilled water. (After dilution to 1 litre, this gives 2.5 mM HEPES.) Add other solutions as follows:

Salt	Stock (g l^{-1})	Volume (ml) added to medium		
		A	**B**	**C**
KH_2PO_4	4.39	1.0	1.0	0.2
$MgSO_4.7H_2O$	25.00	2.0	2.0	2.0
$Ca(NO_3)_2.4H_2O$	40.00		1.0	2.0
NH_4Cl	3.53		1.6	2.2
NaCl	5.00	1.0	1.0	1.0
KCl	5.00	1.0	1.0	1.0
$NaHCO_3$	15.85	1.0	1.0	1.0
$Na_2SiO_3:5H_2O$	43.50	0.25	0.25	0.25
Fe-EDTA stock		0.5	0.25	0.25
Trace metal stock		0.5	0.5	0.5

Buffering

As most soil cyanobacteria can apparently grow over a relatively wide pH range, in most circumstances it is suitable to buffer the medium to pH 7.0. Add HEPES to provide a final concentration of 2.5 mM (= 0.6 g l^{-1}) and adjust to pH 7.0 with

0.1 M NaOH. If the soil is known to have a pH well removed from 7.0, it is recommended to buffer the medium to a value closer to the natural pH. HEPES may be used as the buffer over the range pH 6.5-8.0. (Note that Tris should **not** be used as replacement for HEPES, as Tris is toxic to some cyanobacteria at concentrations tolerated by many eukaryotes.)

Agar plates

Make up 1.5% (w/v) bacteriological grade agar stocks with these various media and pour into Petri dishes to provide a thick agar layer. An 85 mm diam plate requires about 40 ml agar (i.e. thicker layer than used for routine bacteriological practice). The standard protocol requires 48 plates for each of the three media.

Preparation of Inocula

Dissolve 5 mg cycloheximide in 100 ml distilled water, add 1 g soil and shake well. However, unless there is a need to make a quantitative study, it is useful to retain a few small lumps of soil. (This will favour heterogeneity of the environment on the agar plate and perhaps provide local microenvironments more like the natural condition.) Use this suspension to prepare a dilution series in each of the three media **A-C**, diluting by a factor of five each time. Use five dilutions in all (i.e. if the suspension was homogeneous, the final concentration would be 1/5 to 1/3125 of the original. Prepare 100 ml of each dilution.

Inoculation

Use the suspension to inoculate the agar plates. The general procedure is as follows. Spread 0.5 ml suspension in medium **A** on the surface of a medium **A** agar plate (preferably with the aid of a bacteriological spreader) and then use each of the medium **A** dilution series as inocula for other plates. Prepare eight sets of the dilution series for medium **A**. Repeat the procedure for media **B** and **C**, using eight sets of the dilution series for each. Make small insertions with a mounted needle into the surface of the agar for the two highest concentrations of inoculum, in order to encourage growth of filamentous forms growing inside rather than on the surface of the agar. (Note that the volume of inoculum is more than that used in routine microbiological practice; this is to provide a temporary liquid film to encourage growth of dormant organisms.)

Growth Conditions

General facilities

The four sets of environmental conditions refer to different combinations of temperature and light flux. Ideally, this should be achieved using growth rooms with an overhead light source, but, if these are not available, use room temperature and daylight modified to give conditions as similar as possible to those described below. Try to avoid growth conditions likely to lead to rapid drying of the agar, such as a growth room with strong air movements.

Temperature

For three of the four sets, use a temperature towards the upper end of the range likely to be encountered in the field (20 or 25°C for most temperate soils; 30°C for subtropical and tropical soils). The fourth set should be incubated at a lower temperature: 10 or 15°C for temperate soils; 15 or 20°C for subtropical soils; 25°C for tropical soils.

Light flux

At the upper temperature, one set should be incubated under each of the following light conditions (at the surface of the Petri dish):

white *ca.* 150 µmol photon m^{-1} s^{-1} PAR (photosynthetically active radiation);
white *ca.* 15 µmol photon m^{-1} s^{-1} PAR; and
green *ca.* 25 µmol photon m^{-1} s^{-1} PAR.

At the lower temperature, one set should be incubated under:

white light *ca.* 25 µmol photon m^{-1} s^{-1} PAR.

The various conditions can be achieved using sheets of neutral density or green plastic filters.

The flux values are merely intended as a guide. In most cases it seems to be relatively unimportant whether continuous light or a light:dark cycle are provided. However, it can be helpful to provide a light gradient over individual Petri dishes or use different thicknesses of green filter over different parts of one dish.

Incubation

The agar plates should be incubated with the moist surface of the agar upwards. After one to two days, when the liquid on the agar has gone, partially tape the sides of the Petri dish to reduce evaporation.

Possible Modifications to the Protocol

Simplification

If there is a need to study a range of soils, it may be necessary to reduce the number of Petri dishes used for the standard protocol. Omission of the following is the least likely to lead to a loss in observed diversity:

- final (or perhaps even the final two) dilutions
- use of replicates
- incubation at lower temperature.

Any reduction in the combinations of nutrient medium and light flux and quality at the higher temperature poses a serious risk of reducing the number of species detected.

Other Environments

Calcareous and heavy metal-enriched soils

It is recommended that two extra sets of plates are prepared, in which small lumps of surface soil are embedded in the surface of the agar. Where there are lumps of limestone or chalk present in the soil, these should be examined for endolithic growths at the beginning and end of the incubation period.

Saline soils

Add NaCl to the standard media to provide salinity conditions found in nature.

Marine sediments

If well-developed microbial mats are present, the best procedure to establish the range of organisms present is detailed microscopy of a live sample of the mat. (Such mats often show marked gradients of O_2 and E_h which it is difficult to simulate in the laboratory.) Sediments with little, if any, visually obvious surface growths may be treated by a similar protocol to that for soil, but replacing distilled water with filtered seawater or artificial seawater. It is also worth testing combinations of agar media and sediments, with the sediments added shortly before the agar solidifies. It is then necessary to provide live inocula by adding further sediment on the surface and pushed into the agar with a needle.

Limitations of Existing Techniques

Culture conditions on a Petri dish or in liquid culture may not permit some isolates to show their full morphological range necessary for identification. This may sometimes be overcome by subculturing to other combinations of environmental features, but these are not always easy to simulate in the laboratory. It is particularly difficult to provide the right combination of light flux and E_h for studies of intertidal mat communities.

Suggestions to Overcome the Limitations of Techniques

Cyanobacterial studies require a modern flora which brings together the earlier literature. At the same time, the development of specific gene probes would be of great value for complex communities, such as those which occur in rice-fields and marine mats.

Bibliography

Borowitzka, L.A. (1986) Osmoregulation in blue-green algae. *Progress in Phycological Research* 4, 243-256. [A useful review for anyone planning to study cyanobacteria in saline environments.]

Castenholz, R.W. and Waterbury, J.B. (1989) Cyanobacteria. In: *Bergey's Manual of Systematic Bacteriology, Volume 3*. Williams and Wilkins, Baltimore, pp. 1710-1727. [A well-written summary of the state of knowledge at the time of publication.]

Desikachary, T.V. (1959) *Cyanophyta*. Indian Council of Agricultural Research, New Delhi. [The only general botanical flora in English, but unfortunately never reprinted.]

Whitton, B.A. (1992) Diversity, ecology and taxonomy of the cyanobacteria. In: Carr, N.G. and Mann, N. (eds) *Photosynthetic Prokaryotes*. Plenum Press, New York, pp. 1-51. [This includes the original sources for the information quoted without a reference in the present account.]

Whitton, B.A. (1993) Cyanobacteria and *Azolla*. In: Jones, D.G. (ed.) *Exploitation of Microorganisms*. Chapman & Hall, London, pp. 137-167. [Includes many references to studies of soil cyanobacteria.]

Analysis of Non-Culturable Bacteria 5

JAMES M. TIEDJE AND JIZHONG ZHOU

Center for Microbial Ecology, Michigan State University, East Lansing, MI 48824, USA.

The Importance of Non-Culturable Bacteria in Ecosystems

It has been difficult to assess the importance and role of non-culturable microorganisms in ecosystems, since it is usually through culturing that phenotype, and hence role in the environment, can be inferred. However, since non-culturable forms predominate in soil and sediment habitats, usually exceeding culturable organisms by a factor of 10 to 100, it can be reasoned that non-culturable organisms must have a significant role for this large biomass to have been sustained during evolutionary history. Most of these organisms are likely to be heterotrophs, and, hence, must be important in the carbon cycle at least. It is debatable whether the non-culturable organisms are simply dormant cells which do not contribute to the metabolic processes of the habitat, although this issue remains largely unresolved. There certainly are examples of new organisms that have recently been discovered by improved culturing efforts, and, once discovered, their active role in the habitat has been established. Within this large unculturable group, it is likely that a significant number, once recognized, will be shown to have an important function in the environment.

A second potential importance for unculturable organisms is their value to biotechnology. Discovery of novel diversity is the foundation for new biological products, whether they be derived from novel genes or organisms. Pharmaceuticals, enzymes and biocontrol agents are major product areas for which the discovery of new microorganisms is particularly valuable.

Non-cultured microorganisms may also play important roles in habitats by virtue of their interactions with other organisms. Some of these are highly developed symbioses with higher organisms (e.g. plants, animals, insects), whereas others may be loosely linked relationships, such as cross-feeding of unique growth factors. In any case, such relationships are of value to the maintenance of the community and its functions. Symbiotic organisms are often more difficult to culture, as they are highly specialized for the conditions of that association. Therefore, symbionts could be expected to constitute an important fraction of the non-culturable microorganisms.

While the ribosomal approach to the analysis of non-culturable organisms currently offers the greatest potential, other approaches should not be forgotten, as they may have advantages for particular groups of organisms, or may benefit from other technological advances, making them more suitable for use in the future. These other techniques include:

- microscopy-based analyses, such as fluorescent antibody (Wright, 1994) and image analysis techniques (Dubuisson *et al.*, 1994), especially when coupled with the advances in new dye chemistries, lasers and time-resolve capabilities;
- chemotaxonomic approaches, coupled with new analytical methods which improve resolution, sensitivity and structure determination; and
- new culture strategies using multiple gradients (Emerson *et al.*, 1993) which should make some organisms in this group culturable.

Analysis of Soil Samples

The nucleic acid techniques applied to SSU (small sub-unit) rRNA (ribosomal RNA) analysis have created exciting possibilities for rapid detection, identification and characterization of populations, especially of non-culturable bacteria, in environments (Olsen *et al.*, 1986; Pace *et al.*, 1985; Woese, 1987; Ward *et al.,* 1990). These techniques generally include total DNA (or RNA) extraction and purification, PCR (polymerase chain reaction) amplification, gene cloning, DNA sequencing, phylogenetic analysis and gene probe development. This approach has been primarily applied to non-soil environments and there are only a few good examples of its use in soil (Liesack and Stackebrandt, 1992; Stackebrandt *et al.*, 1993). The soil is perhaps the most difficult environment to use molecular techniques, because of the difficulty of comprehensive cell lysis, DNA shearing, humic acid contamination and the extensive bacterial diversity that seems to be present in soil. The steps of analysis are outlined in Fig. 5.1 and described below. Detailed protocols for phylogenetic analysis can be found elsewhere, e.g. Stackebrandt and Rainey (1995).

DNA Extraction and Purification

Two basic approaches can be used for the recovery of DNA from environmental samples such as soils and sediments. The first approach is *indirect extraction*, in

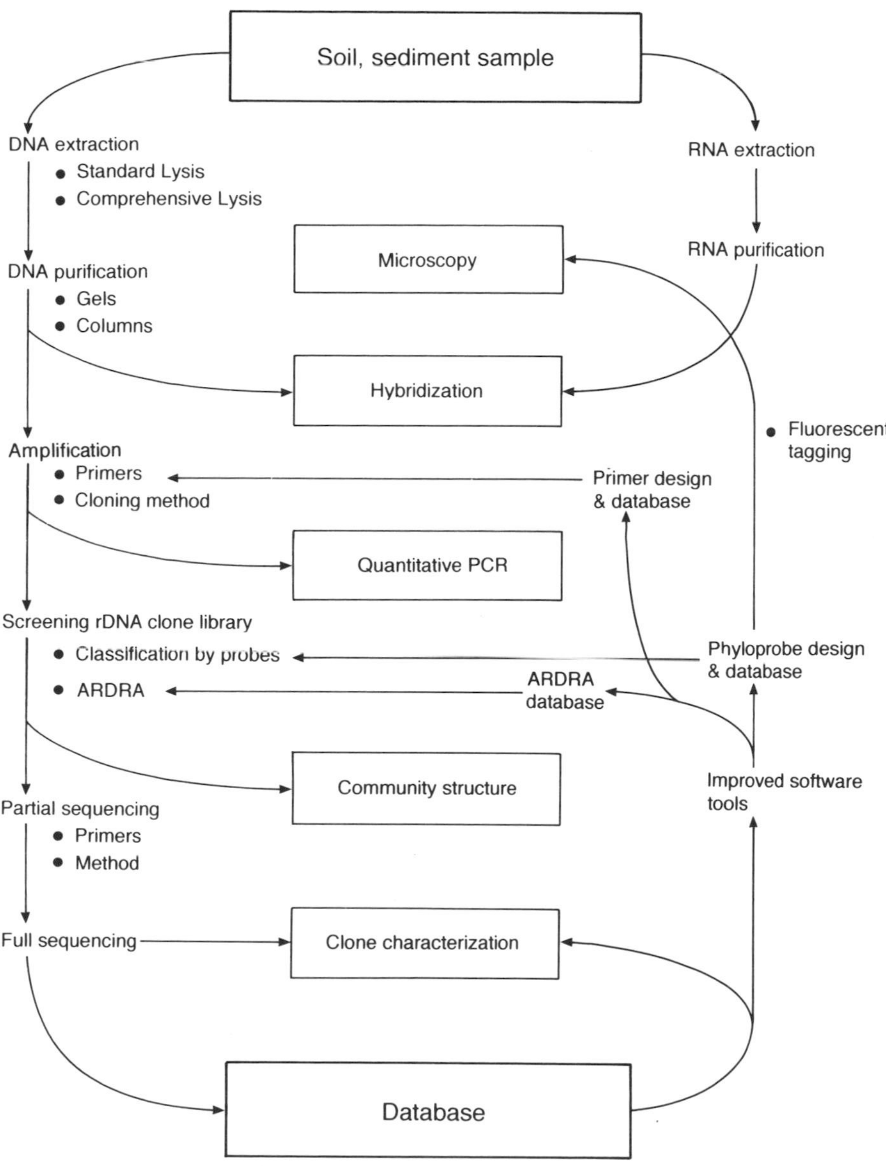

Fig. 5.1. Steps in the SSU rD/RNA analysis of communities. Approaches for characterization, quantitation and community structure analyses are shown in the centre boxes. These are dependent on an efficient sequential analysis of the DNA (scheme on left side) anchored and an improved database (bottom) and probes and primers (right) to analyse the samples. The bulleted items are examples of methodological choices that must be made.

which the microorganisms are first separated by repeated differential centrifugation from most of the soil material before lysis. This method typically recovers less than 50% of the bacterial cells present (Holben *et al.*, 1988). The proportion of cells recovered depends heavily upon soil type. The second approach is *direct extraction*, in which microbes are lysed in the soil and the free DNA is extracted and purified (Ogram *et al.*, 1987). Higher yields of DNA are usually obtained by direct extraction of total DNA, but it always results in co-extraction of other soil components, the most problematic of which are humic substances. Since higher yields can be obtained, direct extraction methods are used more often than indirect methods.

The most commonly used direct extraction method is a high-temperature lysis using SDS (sodium dodecyl sulphate). For example, we have developed a modification of this method which is effective for a variety of soils varying in organic matter and clay contents and pH (Zhou *et al.*, 1996). Soil or sediment samples (5-10 g) are mixed with 13.5 ml DNA extraction buffer (100 mM Tris-HCl (pH 8.0), 100 mM Na_2EDTA (pH 8.0), 100 mM phosphate (pH 8.0), 1.5 M NaCl and 1% CTAB (hexadecyltrimethyl ammonium bromide)) and 100 μl of proteinase K (10 mg ml^{-1}) and shaken at 225 rpm for 30 min at 37°C. After 1.5 ml of 20% (w/v) SDS (sodium dodecyl sulphate) is added, the samples are incubated in a 65°C water bath for 2 h with gentle end-over-end inversions every 15-20 min. The supernatants are collected after centrifugation at 6000 *g* for 10 min at room temperature, transferred into 50 ml capacity centrifuge tubes and extracted with equal volumes of chloroform and isoamylalcohol (24:1, v/v). The aqueous phase is recovered by centrifugation, as described, and precipitated with 0.6 volumes of isopropanol at room temperature for 1 h. A pellet of crude nucleic acids is obtained by centrifugation at 16,000 *g* for 20 min at room temperature, washed with cold 70% ethanol and resuspended in double-distilled sterile water, giving a final volume of 500 μl.

Direct extraction of total DNA always results in co-extraction of other soil components, mainly humic substances, which negatively interfere with subsequent PCR amplification, DNA hybridization and/or restriction enzyme digestion. Therefore, crude DNA extracts must be purified before use. A variety of methods can be used to purify the crude DNA, such as caesium chloride-ethidium bromide density centrifugation, gel electrophoresis and different columns. We have compared several rapid and simple methods to purify crude DNA extracts for speed, cost, DNA purity, molecular weight, PCR amplification, restriction digestion and hybridization (Zhou *et al.*, 1996). The purification method which is most generally useful is to combine agarose gel electrophoresis and commercially available resins. The DNA is first electrophoresed on agarose gel and then recovered using 'Wizard PCR Preps' purification resin (Promega, Madison, USA). The DNA is then pure enough for PCR amplification, DNA hybridization and restriction enzyme digestion.

PCR Amplification of the 16S rRNA Gene

The 16S rRNA gene is the key molecule for establishing bacterial phylogeny. These genes can be directly amplified from the purified environmental DNA samples using two conserved primers which bind to both ends of the 16S rRNA gene. The most commonly used eubacterial amplification primers are:

5'-AGAGTTTGATCCTGGCTCAG-3' and
5'-CGGTTACCTTGTTACGACTT-3' (Weisburg *et al.*, 1991).

The 16S rRNA gene can be amplified in a 100 μl reaction containing 1 × *Taq* polymerase buffer, 2.5 units *Taq* polymerase, 20 mol dNTPs and 100 pmol of each primer. The reaction conditions consist of denaturation at 92°C for 2 min, 30 cycles of 94°C for 15 s, 55°C for 30 s and 72°C for 2 min, plus one additional cycle with a final 6 min of chain elongation. Samples are stored at 4°C. The amplified 16S rRNA gene fragment, about 1.5 kb, can be visualized by agarose gel electrophoresis.

16S rRNA Gene Cloning and Screening

Cloning of the PCR-amplifed 16S rRNA gene is often difficult and labour-intensive. Several strategies can be used to clone the PCR products amplified from environmental samples, such as blunt-end cloning, the sticky-end cloning which is facilitated by introducing polylinker tails to the primers containing restriction endonuclease sites, and TA-cloning, which is based on the template-independent deoxylnucleotidyl activity of *Taq* DNA polymerase. The first two strategies require several steps to modify the PCR-amplified products, which may result in an additional decrease of cloning efficiency and increase of cloning bias. The TA-cloning approach, based on the template-independent deoxylnucleotidyl activity of the *Taq* DNA polymerase, is straightforward. We have evaluated the applicability of the TA-cloning approach to microbial community studies for cloning efficiency and bias. The amplified 16S rRNA genes are ligated with the pCRTM II vector from Invitrogen (San Diego, USA) and transformed into *Escherichia coli* cells. The TA-cloning approach is efficient and appears to have little cloning bias resulting from plasmid ligation, plasmid transformation and template heterogeneity.

Screening of the cloned PCR-amplified 16S rRNA genes is one of the major tasks in characterizing microbial communities using molecular tools. We have developed a simple method for rapid screening of the cloned genes without involving plasmid isolation. The cloned 16S rRNA gene inserts are directly amplified from transformant cells using vector-specific primers. This method substantially reduces the time and cost of cloning PCR-amplified 16S rRNA genes.

Two major approaches can be used to detect unique 16S rRNA gene clones. RFLP (restriction fragment length polymorphism) analysis (Moyer *et al.*, 1994) and single nucleotide sequence track analysis. The former is used more often than the latter. In RFLP analysis, the amplified inserts using the vector-specific primers can be restricted using each of the tetrameric endonuclease pairs *Msp* I plus *Rsa* I, and *Hha* I plus *Hae* III incubated at 37°C overnight. The resulting RFLP products can be separated by gel electrophoresis in 3.5% Metaphor agarose in 1 × TBE at 4°C with 7 volts cm^{-1} for 4 h. The RFLP patterns can be compared by eye or by using computer software.

Determination of Nucleotide Sequences

The DNA sequence of the cloned 16S rRNA gene can be manually or automatically determined by the dideoxy chain termination method, using T7 DNA polymerase or *Taq* DNA polymerase. The advantage of using *Taq* polymerase is that the sequencing reaction can be carried out at high temperature, so that interference resulting from the secondary structure of the templates on the sequencing reaction will be alleviated.

Many different commercially available kits can be used for manual sequencing. For example, DNA sequences can be directly determined from PCR-amplified DNA as templates, using the double-stranded DNA cycle sequencing system from BRL (Gaithersburg, USA). DNA sequences can also be determined with automated fluorescent *Taq* cycle sequencing, for example, using the ABI Catalyst 800 and ABI 373A sequencer (Applied Biosystems, Foster City, USA).

Since the 16S rRNA gene consists of a mosaic structure of regions with varying degrees of conservation, sequencing primers can be synthesized based on the sequences of the conserved regions. At least 10 primers are needed for obtaining nearly the full-length 16S rRNA gene sequence from both directions. According to *Escherichia coli* 16S rRNA gene positions, the internal forward sequencing primers used in most studies are 342-357, 519-533, 787-802 and 1099-1114. The internal reverse primers are 357-342, 529-515, 802-787 and 1115-1100. In addition, the forward and reverse primers for PCR amplification of the 16S rRNA genes can also be used to determine the sequences of both ends of the cloned 16S rRNA gene. These sequencing primers work very well for the majority of eubacteria. Other primer sets are needed for *Archaea* and novel, deeply rooted organisms.

Phylogenetic Analysis

DNA sequences from each sequencing reaction can be assembled using different computer programs to obtain the full length of 16S rRNA gene sequence, e.g. the gel assembly programs in the Genetics Computer Group (GCG) software package. Since 16S rRNA genes differ significantly in length, they are generally aligned by hand using a number of prealigned sequences as a reference. The

Ribosomal Database Project (RDP) provides such prealigned sequence format for most 16S rRNA genes from prokaryotes. Different sequence editing and analysis programs, such as GDE, AE2, SUBALIGN, READSEQ and CONVERT-ALAN are also available in RDP.

Three major types of tree-inferring methods can be used for phylogenetic analysis: DNA distance matrix, maximum parsimony and maximum likelihood. Each method has different advantages and disadvantages. For example, distance methods are very fast and suitable for a large number of data sets, but they do not consider the evolutionary information about individual positions. While maximum likelihood methods consider the evolutionary information on a site by site basis, and do not require constant evolutionary rates among different lineages, they are very computationally intensive. Generally, all three methods should be used, as the strength of the phylogenetic inference is enhanced if additional methods to produce trees give basically the same branching pattern. Of the three, the maximum likelihood method gives more robust results.

Phylogenetic Gene Probe Development

Phylogenetic gene probes can be developed based on 16S rRNA gene sequences (Amann *et al.,* 1995). Once the phylogenetic relationships of interesting organisms are known, PCR primers or hybridization-based oligonucleotide probes specific to these organisms may be designed by comparing sequence alignments. The probe specificity should be tested by using closely related organisms as references. The developed probes can be used for monitoring organism groups of interest in the environment. In addition, fluorescein-labelled oligonucleotides (riboprobes) can also be developed to hybridize to cell ribosomes to allow microscopic enumeration of cells in the sample matrix (Stahl and Amann, 1991).

Limitations of Existing Techniques

While the approach of using 16S rRNA sequence information to determine evolutionary relatedness among bacteria and to aid taxonomic identification has been a powerful advancement for microbial diversity studies, there remain a number of significant limitations to its routine use. Some of these are typical of any technology in its infancy, such as lack of automation, high cost and the need for specialized expertise. However, continued interest usually leads to further developments which reduce these barriers. We identify here the major issues which are delaying widespread access to this technology and other technical barriers to its efficient use in soil and sediment ecology studies.

Time Constraints

At the moment, the protocol is lengthy and most steps must be done by hand, as no automated procedure has yet been developed. This limits the capacity for the number of samples which can be handled and increases personnel costs. The time-consuming steps are DNA extraction/cell lysis, DNA clean-up and (especially) screening of the clone library for classes of rRNA genes. In a typical sample, at least 100 clones should be screened to give a reasonable representation of the community structure (the 'species' richness pattern), and to enhance the chance for recovering ribosomal genes from more novel, but rare, organisms. It currently takes 10-14 days by a skilled scientist with all facilities readily available to complete the analysis of one soil or sediment sample with ≥ 100 clones to the stage of sequencing. This constraint clearly limits the number of samples that can be analysed as well as the statistical analyses which can be performed.

Cost

The materials (particularly enzymes) and equipment are relatively expensive, especially compared with the cost levels that the ecology and systematics communities are traditionally accustomed to. The final steps of sequencing and database analysis are even more costly in terms of both equipment and specialized expertise. This cost factor certainly makes it unfeasible to have numerous independent and complete laboratories using this approach around the world. The more likely approach is to have broad-based research teams with the costly steps done at centralized facilities.

Bias

The problem of bias in the molecular approach, i.e. that recovered clones do not reflecting the natural organism distribution, has been much discussed. Bias can occur at most stages of the analysis and, at our current level of knowledge, must be assumed to occur to some extent. But advances are occurring regularly in methods, such as more comprehensive cell lysis, improved primers and cloning methods, which reduce bias, or which, at least, allow the degree of bias to be estimated. Since ribosomal sequences of new organisms are being discovered at a rapid pace, this method is clearly enhancing our knowledge beyond what was known from culture-based methods. Bias is not so severe as to make it difficult to discover new microbial diversity. Also, when the goal is to discover novel diversity, the problem of bias is less significant than if a comprehensive and quantitative analysis of microbial community structure and composition is being attempted.

Extensive Diversity

Recent information from soil and sediment environments suggests that the extent of microbial diversity in any sample is extremely high - overwhelming might be a more apt term. This fact, together with the slow speed of analysis, makes any attempt at comprehensive analysis impractical, even in the best equipped laboratories. If thousands of 'species' exist in each 1 to 10 g of sample, which seems likely from current data (Torsvik *et al.,* 1990), then the screening of 100 clones is clearly insufficient to recover total diversity. Therefore, methods to reduce community complexity and focus on particular groups are necessary.

Blindness at the 'Species' Level

Diversity studies require some definition of a unit or units which circumscribes members with like features and recognizes meaningful differences among groups with other features. 'Species' is used as this unit of distinction in biology and microbiologists also use this term, but without a widely accepted means to define or to measure species delimitations. A particular problem for microbial diversity studies is the lack of an efficient method to distinguish organisms at the species level, and so this level of distinction is often avoided. (Note that taxonomic descriptions such as *Pseudomonas* sp. are often seen in the literature.) Furthermore, the 16S rRNA gene is too conserved to be routinely useful for resolution at the species level (Stackebrandt and Rainey, 1995). Differences at the species level represent only a 0-3% difference in 16S rRNA sequence. The use of the 16S rRNA method at present should largely be at the genus level of distinction, with the finer level being reported as sequence similarities, which may or may not translate to useful species designations now or in the future.

Functional Interpretation

The growing interest in ribosomal sequence analysis has led to the criticism that microbial activities are being ignored, especially as it is this activity which justifies interest in microorganisms. But characterization of uncultured organisms based on phylogeny is a sound approach to gain insight into activity and physiology as well, since many of these features are often conserved in evolution. Such information may also provide insight into how to culture an apparently non-culturable organism. Scientists using the ribosomal methodology should make a better effort to attempt to use the phylogenetic information in an ecological or physiological context, so that a more comprehensive insight into the microbial community is fostered, rather than ignored. An important limitation to easy functional interpretation is the lack of an integrated microbial database, so that phenotypic features of related organisms are readily apparent.

Suggestions to Overcome Technical Limitations

Database Improvements

The ribosomal approach is dependent on the ribosomal database for its value, and, for the next generation of use, several improvements are important (Fig. 5.1). These include;

- software to more rapidly update the database with new sequences and to handle the increased rate of growth of data;
- a systematic effort to fill the gaps in the database especially with sequences from organisms that are already well described; and
- development of software to rapidly design and evaluate probes and primers for subgroups of environmentally important organisms.

An effort to complete the database with at least the well-described type strains is the most important goal, so that new sequences can be compared with those already known. An equally important part of this goal is to enhance the number of environmental sequences in the database. With the addition of both well-described strains and environmental sequences, new primers and probes may be developed more specifically for important environmental groups, including non-culturable organisms.

The second goal is to integrate ribosomal data with phenotypic data especially, but also with chemotaxonomic, nomenclatural, ecological, metabolic, sequence and image information data in a way that supports complex queries and seamless navigation. This is critical to bringing physiological meaning to sequence information, as it makes the connection nearly automatic. An additional advantage of this linkage is that it will help advance the definition of the prokaryotic species, since the polyphasic information will be integrated and in database format, enabling quantitative comparison. The database efforts are listed first since it is the foundation on which a large-scale programme in biodiversity must be built. Microbial biodiversity studies cannot efficiently proceed without attention to the database; the key role of the database is also apparent from Fig. 5.1.

Speed, Cost and Extensiveness of Surveying Microbial Diversity

The screening of ribosomal clones and the extensive microbial diversity in these habitats cause the ribosomal approach to be slow and costly for even a cursory evaluation of dominant members. The keys to overcoming these limitations lie in developing probe systems to screen clones, and in developing primers for particular subgroups to reduce the complexity of the community under study (Fig. 5.1). These are dependent on advances in the database and its software tools. A related technological need is for a convenient, inexpensive, perhaps even disposable, 'plate' with a bank of 100 or more specific oligonucleotide probes for fast screening of ribosomal clone libraries, as well as to quantify the different members of the community by hybridization (Fig. 5.1).

Overcoming Bias

Two approaches are being used and both are on track to address this problem. The first is the development of improved procedures for cell lysis, improved primers and improved cloning methods. The second is to conduct studies with defined mixtures of organisms of different phylogenies to evaluate the extent of bias. The latter approach needs more robust tests on environmentally important organisms.

Integration of Techniques into a General Scheme of Analysis

The ribosomal analysis is one line of several complementary analyses important to microbial biodiversity studies (Fig. 5.2).

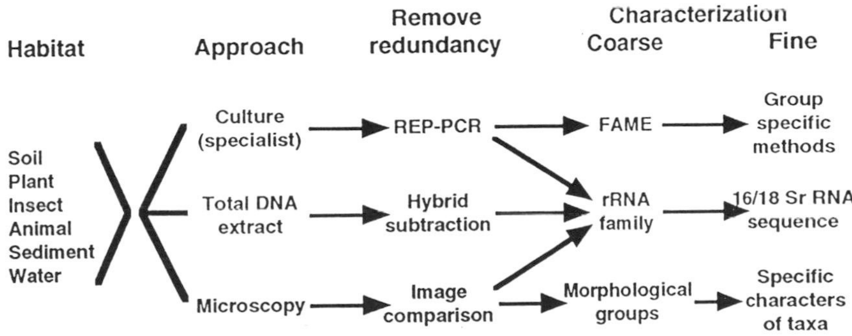

Fig. 5.2. General scheme for total microbial diversity analysis following three different complementary approaches.

If these are carried out in a co-ordinated way, the interaction among the types of analyses can greatly advance the interpretation. The scheme illustrated employs a hierarchical approach in which redundant organisms are first removed from the analysis scheme as efficiently as possible, followed by a coarse level of characterization, and then a decision is made about which organisms warrant a finer level of characterization. In the case of ribosomal analyses, the coarse level is usually obtained by a partial sequence and the finer level by sequencing the entire 16S rRNA gene.

Bibliography

Amann, R.I., Ludwig, W. and Schleifer, K.H. (1995) Phylogenetic identification and *in situ* detection of individual microbial cells without cultivation. *Microbiological Reviews* 59, 143.

Dubuisson, M.P., Jain, A.K. and Jain, M.K. (1994) Segmentation and classification of bacterial culture images. *Journal of Microbiological Methods* 19, 279-295.

Emerson, D., Worden, R.M. and Breznak, J.A. (1993) A diffusion gradient chamber for studying microbial behavior and separating microorganisms. *Applied and Environmental Microbiology* 60, 1269-1278.

Holben, W.E., Jansson, J.K., Chelm, B.K. and Tiedje, J.M. (1988) DNA probe method for the detection of specific microorganisms in the soil bacterial community. *Applied and Environmental Microbiology* 54, 703-711.

Liesack, W. and Stackebrandt, E. (1992) Occurrence of novel groups of the domain Bacteria as revealed by analysis of genetic material isolated from an Australian terrestrial environment. *Journal of Bacteriology* 174, 2072-2078.

Moyer, C.L., Dobbs, F.C. and Karl, D.M. (1994) Estimation of diversity and community structure through restriction fragment length polymorphism distribution analysis of bacterial 16S rRNA genes from a microbial mat at an active, hydrothermal vent system, Loihi Seamount, Hawaii. *Applied and Environmental Microbiology* 60, 871-879.

Ogram, A., Sayler, G.S. and Barkay, T. (1987) The extraction and purification of microbial DNA from sediments. *Journal of Microbiological Methods* 7, 57-66.

Olsen, G.J., Lane, D.J., Giovannoni, S.J. and Pace, N.R. (1986) Microbial ecology and evolution: a ribosomal RNA approach. *Annual Review of Microbiol*ogy 40, 337-355.

Pace, N.R., Stahl, D.A., Lane, D.J. and Olsen, G.J. (1985) The analysis of natural microbial communities by ribosomal RNA sequences. *Microbiology Ecology* 9, 1-56.

Stackebrandt, E. and Rainey, F.A. (1995) Partial and complete 16S rDNA sequences, their use in generation of 16S rDNA phylogenetic trees and their implications in molecular ecological studies. In: Akkermans, A.D.L., van Elsas, J.D. and de Bruijn, F.J. (eds) *Molecular Microbial Ecology Manual.* Kluwer Academic Publishers, Dordrecht, pp. 1-17.

Stackebrandt, E., Liesack, W. and Goebel, B.M. (1993) Bacterial diversity in a soil sample from a subtropical Australian environment as determined by 16S rDNA analysis. *FASEB Journal* 7, 232-236.

Stahl, D.A. and Amann, R.I. (1991) Development and application of nucleic acid probes. In: Stackebrandt, E. and Goodfellow, M. (eds) *Nucleic Acid Techniques in Bacterial Systematics.* John Wiley and Sons, Chichester, pp. 205-248.

Torsvik, V., Goksoyr, J. and Daae, F.L. (1990) High diversity in DNA of soil bacteria. *Applied and Environmental Microbiology* 56, 782-787.

Ward, D.M., Weller, R. and Bateson, M.M. (1990) 16S rRNA sequences reveal numerous uncultured inhabitants in a natural community. *Nature* 345, 63-65.

Weisburg, W.W., Barns, S.M., Pelletier, D.A. and Lane, D.J. (1991) 16S ribosomal DNA amplification for phylogenetic study. *Journal of Bacteriology* 173, 697-703.

Woese, C.R. (1987) Bacterial evolution. *Microbiological Reviews* 51, 221-271.

Wright, S.F. (1994) Serology and conjugation of antibodies. In: Weaver, R.W., Angle, J.S. and Bottomley, P.S. (eds) *Methods of Soil Analysis. Part 2. Microbiological and Biochemical Properties.* Soil Science Society of America, Madison, pp. 593-618.

Zhou, J., Bruns, M.A. and Tiedje, J.M. (1996) DNA recovery from soils of diverse composition. *Applied and Environmental Microbiology* (in press).

Algae 6

JOHN D. DODGE[1] AND L. ELLIOT SHUBERT[2]

[1]Biology Department, Royal Holloway, University of London, Egham, Surrey TW20 0EX, UK; [2]Department of Botany, The Natural History Museum, Cromwell Road, London SW7 5BD, UK.

Algae in Terrestrial and Aquatic Ecosystems

Algae are the primary producers of below-ground ecosystems and they also play an important role in soil genesis and successional processes, which affect physical and chemical amelioration. The algae are essential components of grassland soils, desert rocks, hot and cold desert soils, tropical and temperate forest soils, agricultural soils, wet soils (silt and mud), particularly those of paddy fields and marine marginal situations such as mud flats, salt marshes and sandy beaches.

They are also present in the sediments of shallow freshwater ponds, lake margins and bottoms and seas, both estuarine and of deeper water. In deeper freshwater and marine sediments, they are generally found as resistant bodies,enabling long-term survival, rather than as vegetative cells. In addition to the resistant stages all sediments contain algal debris such as silica scales, skeletal or wall structures, and calcareous scales. These, representing the previous occurrence of members of the *Bacillariophyta, Chrysophyta* and *Haptophyta*, in the sediment or supernatant water, can be of great value to biosystematists and in studies of palaeoecology. An outline of the main groups of algae in these various habitats is illustrated in Fig. 6.1.

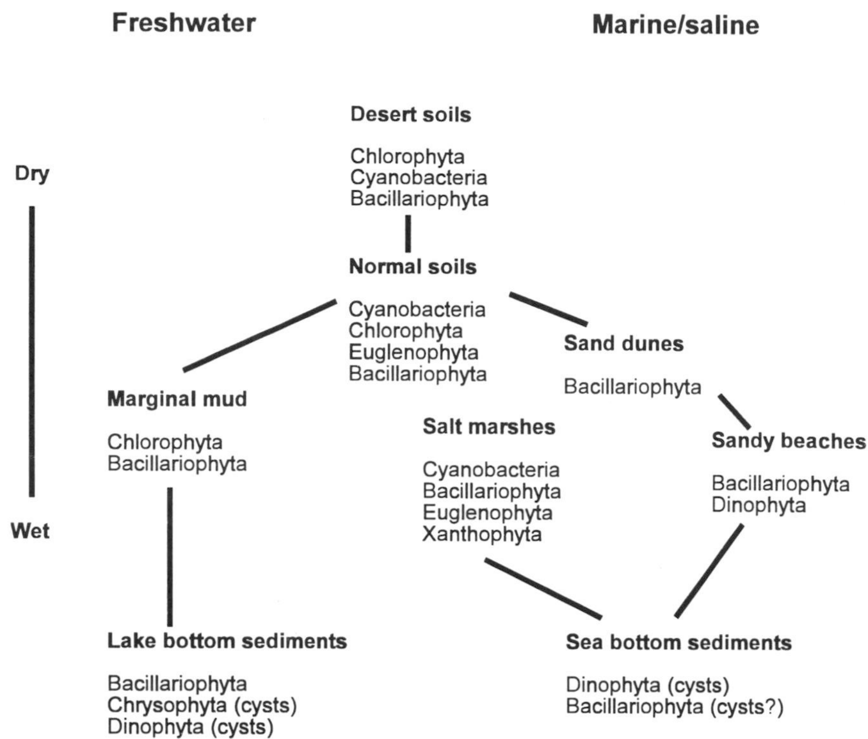

Fig. 6.1. Soil and sediment types where algae are found. The main groups of algae present are noted in each case.

Sampling Methods

Soils and Surfaces

Use a sterile disposable plastic spatula or a garden trowel (surface-sterilized with 70% (v/v) ethanol) to scoop a surface sample of soil. Alternatively, use a corer (Colijn and Dijkema, 1981) or a container with a known area pressed into the surface to obtain a standard sample (reviewed in Eaton and Moss, 1966). To collect at various soil depths or soil horizons, it is necessary to dig a hole and take samples aseptically from the exposed lateral soil face/layers. Several scoops of soil should be collected in a sterile plastic container, such as a 'Whirlpak' bag or 'Ziplock' bag, and labelled with the date and place of collection.

Beaches and Mud Surfaces

The method is as for soils and surfaces, plus use of motility and phototaxis to collect organisms on a known area of lens tissue placed upon the moist surface (Eaton and Moss, 1966). This method has been successfully used for epipelic desmids, diatoms, euglenoid flagellates (Palmer and Round, 1965) and mixed samples containing diatoms, green flagellates, euglenoids, and blue-green algal filaments (Round and Eaton, 1966). An alternative method which may be used for coarse substrates such as beach sand is to place a portion of substrate in an open tube with silk or nylon netting closing the bottom. This is suspended above a collecting dish and chilled water placed in the top to flush out the microorganisms (Uhlig, 1964). For motile organisms such as diatoms, which move over surfaces, artificial substrates in the form of microscope cover glasses can be placed on the wet mud or sand sample in a Petri dish. After a period these are examined by placing on a drop of water or stain on a microscope slide and the adhering cells identified and counted.

Underwater Sediments

Use of a grab is suitable for crude samples but a coring device is necessary for more precise sampling with retention of the core structure and the flocculant sediment surface. The Jenkin mud sampler (Fig. 6.2) has been devised for freshwater sediments (Ohnstad and Jones, 1982). For marine sediments more robust devices such as the Craib corer (Craib, 1965) or HAPS corer (Kanneworff and Nicolaisen, 1973) have been used to collect dinoflagellate cysts (Lewis *et al.*, 1984; Ellegaard *et al.*, 1994). In both cases these corers penetrate the sediment and then can be closed so that a relatively undisturbed core with its flocculant surface layer can be obtained.

Sediment traps moored at the bottom of the water body, or suspended on a line at a known depth, have been used to provide a known time-scale and quantification to sediment accumulation when the algae drop from the photic zone (Odate and Maita, 1990). Very sophisticated multiple samplers which can be opened for set periods of time have been devised for oceanic situations (Honjo *et al.*, 1980; Honjo and Doherty, 1988).

Storage and Preservation

Soils collected from terrestrial habitats may be stored dried. If the soil was wet at the time of collection, it should be air-dried before storage. A recent study (Trainor and Gladych, 1995) has shown that five taxa of green algae have survived in desiccated soil for 35 years, a decrease of two taxa in the last 10 years. However, there had been a considerable decrease in viable cells from 2000 per gram at 25 years to only 30-50 at 35 years.

Wet samples, such as those from mud surfaces and beaches, should be stored in cool conditions under subdued light when the time between collection and examination is extended for more than 2-3 days. Sediment samples are stored under refrigeration or may be frozen. With marine sediment samples, it has been shown that the viability of dinoflagellate cysts decreased from 97% to 3% over nine years when stored at -5°C in the dark (Lewis *et al.*, 1996). For preservation and long-term storage chemical fixatives such as formaldehyde are used.

Fig. 6.2. Drawing to illustrate the Jenkin surface-mud sampler in position partly embedded in the mud at the bottom of a lake (after Ohnstad and Jones, 1982).

Examination and Extraction Methods

With surface samples the algae are often concentrated enough to be visible when a small aliquot is examined directly by light microscopy. With flagellates such as dinoflagellates in beach samples these are much better observed alive, when the swimming movement of the cells enables them to be spotted amongst the sand grains and detritus. If detailed examination is necessary cells can be picked out by micropipette and placed on another slide for high-power light microscopy or can be fixed and mounted on a filter in order to process for scanning electron microscopy.

Where sediments are required to be examined directly the use of fluoresence microscopy can be employed. The natural red fluorescence of chlorophyll enables living cells to be spotted amongst the mineral and organic debris or the sample can be suspended in a fluorochrome such as DAPI (4',6-diamidino-2-phenylindole), which is only absorbed by the living matter. A specific technique for cysts of the dinoflagellate *Alexandrium* involves the use of the fluorochrome primuline, which gives the cysts an intense yellow-green fluoresence which enables them to be readily enumerated in coastal sediments (Yamaguchi *et al.*, 1995). Immunofluorescence can be used for the direct observation and identification of soil algae using epifluorescent microscopy (Fliermans and Schmidt, 1977).

Suspension of the sample in water or culture medium, allowing the mineral particles to settle out and then concentrating the supernatant can be used to extract the algae from some soils. Alternatively, a soil suspension can be aspirated onto a nutrient agar plate.

Implanted Slide Technique

This is a novel technique for direct examination of indigenous soil algae developed by Pipe and Cullimore (1980). The procedure involves the implantation of a pre-cleaned microscope slide vertically in the soil, leaving the top 1.5 cm of the slide projecting above the soil surface. Light is transmitted from the exposed surface to the buried portion, encouraging algal growth on the slide surface in contact with the soil. After incubation, the slide is removed and examined microscopically.

Culture Techniques

These are used for both soils and sediments and involve inoculation of a liquid or solid (agar) culture medium with either a selected (as in the case of surface-living algae) or homogenized aliquot of the sample followed by leaving the culture vessel in a north window or in a culture cabinet. It is best to try several different types of inorganic mineral media and various light and temperature regimes to cultivate the variety of algae present in the soil. The culture is examined at intervals to see if algae have grown up. Further isolation and subculturing will be necessary if unialgal or pure cultures are required (Wiedeman *et al.*, 1964)

A most probable number (MPN) method (Throndsen, 1978) has been used to obtain quantitative results from culture techniques (Harris *et al.*, 1995; Lewis *et al.*, in press). Here an aliquot of sediment is inoculated into a suitable medium using a range of dilutions. When enumerating the cultures the MPN table is used to derive cell numbers in the original sample.

Cleaning Techniques

For resistant stages in sediments many techniques have been devised, particularly by palynologists. A sieving and settling technique for dinoflagellate cysts (Matsuoka *et al*., 1989) is shown in Fig. 6.3. This involves the use of hydrochloric acid and hydrofluoric acid to dissolve detritus and mineral matter leaving only the sporopollenin-walled dinoflagellate cysts. A density gradient centrifugation method using colloidal silica gel ('Ludox T.M.') is also useful for some types of sediment (Price *et al.,* 1978). A combined palynological and culturing technique for dinocysts that was used successfully by Wall and Dale (1968) is illustrated in Fig. 6.4 (after Dale, 1979).

Where only diatoms are present, or they are the only organisms to be enumerated, a number of techniques are available (Hasle and Fryxell, 1970; Hasle, 1978). These involve the use of strong acids or oxidizing agents to dissolve the cell contents and organic detritus and allow the taxonomically important silica cell wall (frustule) to be examined by light or electron microscopy.

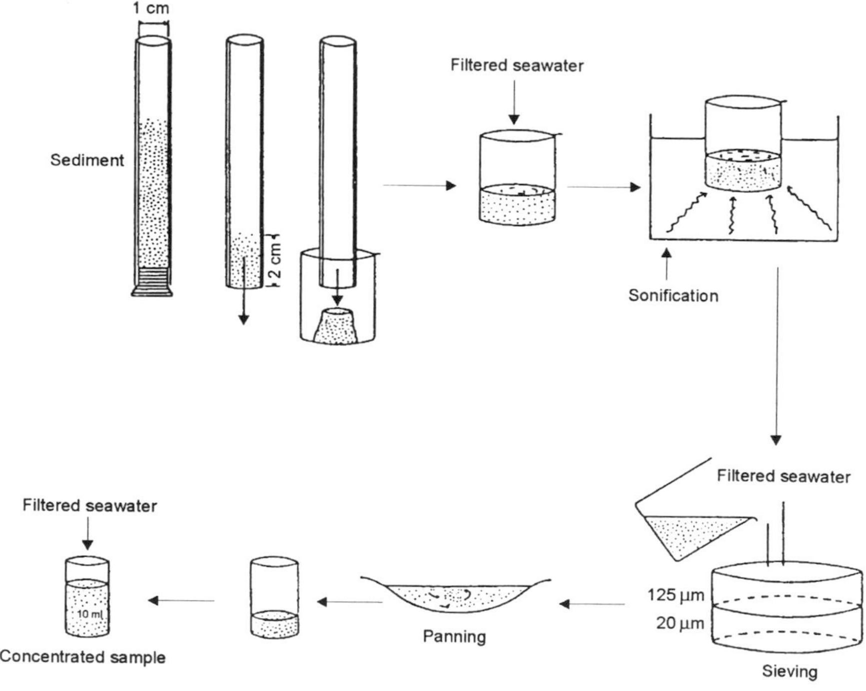

Fig. 6.3. Diagrams to show the sonication method for the extraction and concentration of marine dinoflagellate cysts. (After Matsuoka *et al.*, 1989.)

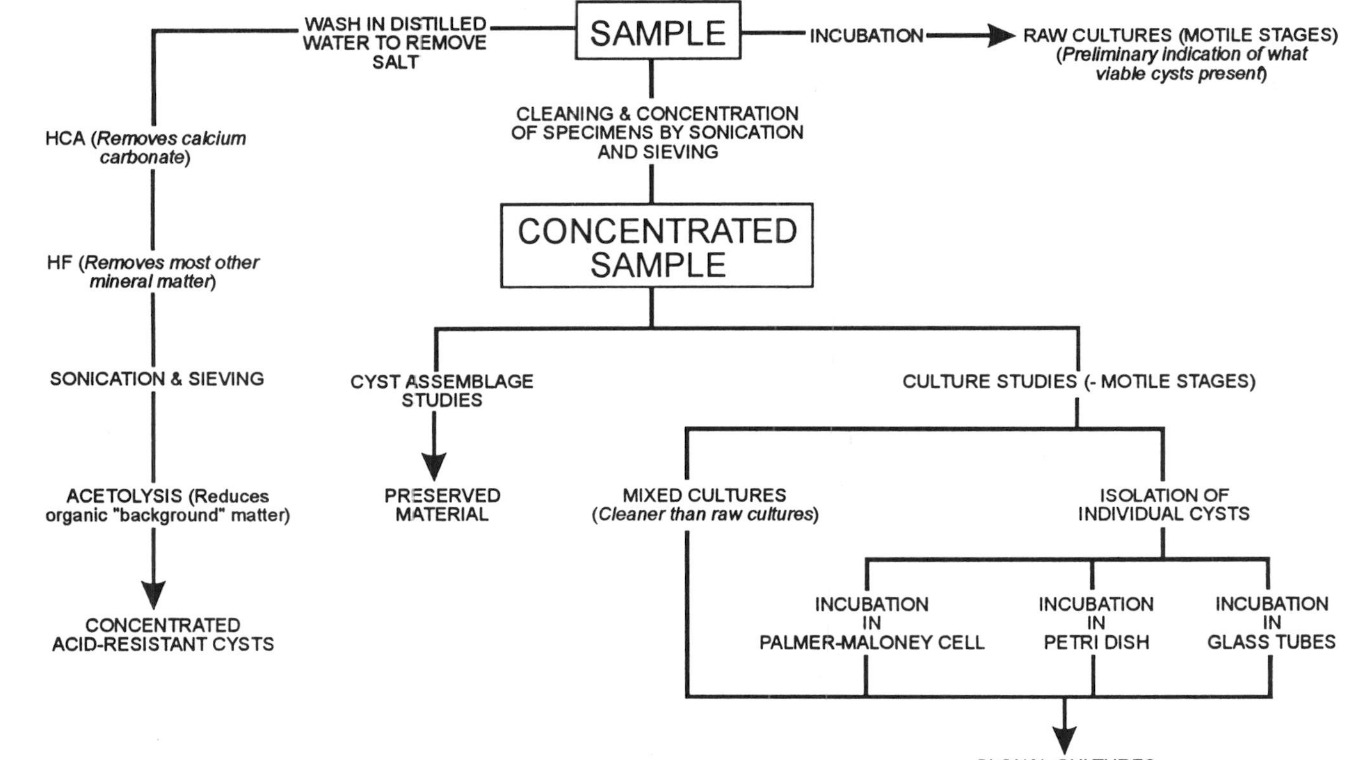

Fig. 6.4. A flow chart to illustrate the combined palynological and culture technique.

Extraction Techniques

Extraction of the chlorophyll *a* pigment from the soil or sediment provides an estimate of total algal biomass (Sharabi and Pramer, 1973; Shubert and Starks, 1979). Chlorophyll *a* can be extracted in subdued light by adding 10 ml dimethyl sulphoxide, 40 ml of reagent grade 90% (v/v) acetone and a pinch of magnesium carbonate to 5 g of soil. The mixture is homogenized in a blender for four minutes and filtered immediately through filter paper (Whatman No. 1). The supernatant is measured on a spectrophotometer, spectrofluorometer or a fluorometer, and the value is corrected for phaeophytin *a* (a degradation product of chlorophyll *a*) (Sharabi and Pramer, 1973; Shubert and Starks, 1979).

Problems with Existing Methods

Sampling suffers from the general problem resulting from the patchy distribution of the organisms; in the case of dinoflagellate cysts in sediments this has been shown to be both vertical and horizontally variable (Anderson *et al.*, 1982; Balch *et al.*, 1983).

Examination of samples suffers obstruction or difficulty of viewing the algae because of mineral and organic matter. The fluorescence techniques mentioned above are a great aid in revealing the organisms in spite of the detritus.

Loss of cells can be a problem during cleaning or extraction processes, particularly when they involve strong chemicals such as palynological methods. Similarly there may be difficulty in culturing resistant stages and unreliability of data since some species are bound to be more easy to culture than others. For this technique, there is uncertainty about the viability of samples with storage.

Problems in converting sample counts into a meaningful population count arise with all of the techniques of collection and enumeration.

One should be aware that the MPN method fails to account for the varying growth rates, morphological forms and nutritional and physical requirements of the algae (Starks *et al.*, 1981). Quantifying species abundance or biovolume is difficult, since there is not a suitable method that accurately takes into account every species and morphological forms. And there is the problem of vegetative cells versus resistant spores, akinetes, etc. Are resistant forms germinating when the soil or sediment is cultured? It is not unusual to find many single cells and/or clumps of green algae which defy traditional taxonomic keys. It most cases it is necessary to subculture these microscopic green algae for a definitive identification.

Improvements to the Methods

Confusion exists in some areas where two types of structure look similar by normal microscopy or where there is an excess of detrital matter. A technique has been developed for rather featureless dinoflagellate cysts. Here, a simple polarized light examination between crossed Nicol prisms highlights dinoflagellate cysts from background particles by a characteristic cross-like interference pattern (Reid and Boalch, 1987).

An improved concept of the range of algae in soils or sediments could be achieved if a satisfactory method was developed that extracted all of the photosynthetic and accessory pigments present which could be quantified spectrophotometrically. The pigment profile could be compared with the known pigment complements of the major algal divisions to assess the relative contributions of the different groups.

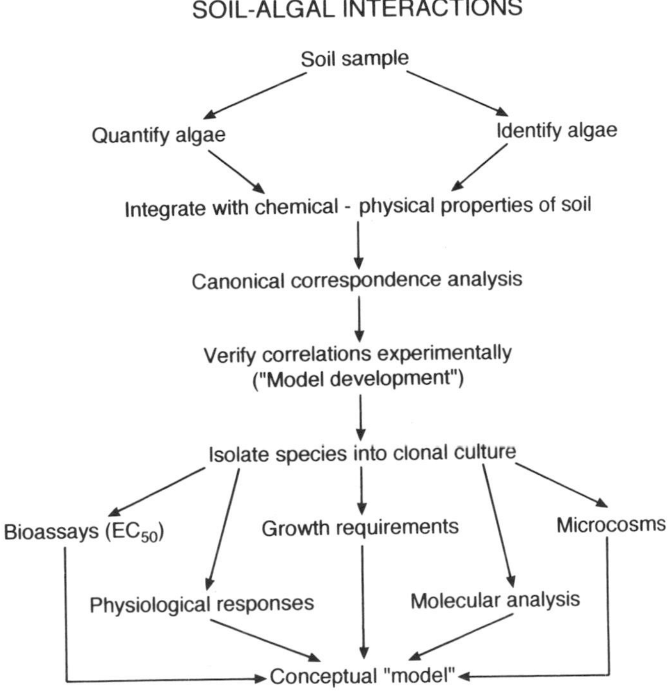

Fig. 6.5. Synopsis of an approach for using soil algae to study biodiversity.

It is essential that a standard protocol be established to study the algae inhabiting the below-ground ecosystem. This will ensure similar collection techniques, species identification and quantifying procedures, and data collected from diverse sites will be comparable. Using a holistic approach to the study of soil algae will improve our understanding of their physiological/biochemical contribution to the soil/sediment ecosystem (Shubert and Starks, 1985). A model concept is illustrated in Fig. 6.5.

Integrating Techniques into a General Scheme of Analysis

A variety of methods must be employed to elucidate the diversity of soil microorganisms, since there are a variety of growth requirements (nutrients, light, temperature). Techniques for heterotrophic microorganisms do not necessarily select for the photoautotrophic organisms. Thus, specialized techniques unique to algae must be used. Soil biodiversity infers soil fertility, since the productivity of a soil is a combination of the physical, chemical and biological factors. Soil and sediment algae are meaningful bioindicators of soil fertility. Biodiversity of soil/sediment ecosystems can be overlooked easily.

Bibliography

Anderson, D.M., Aubrey, D.G., Tyler, M.A. and Coats, D.W. (1982) Vertical and horizontal distributions of dinoflagellate cysts in sediments. *Limnology and Oceanography* 27, 757-765.

Balch, W.M., Reid, P.C. and Surrey-Gent, S.C. (1983) Spatial and temporal variability of dinoflagellate cyst abundance in a tidal estuary. *Canadian Journal of Fisheries and Aquatic Science* 40 (Suppl. 1), 244-261.

Colijn, F. and Dijkema, K.S. (1981) Species composition of benthic diatoms and distribution of chlorophyll *a* on an intertidal flat in the Dutch Wadden Sea. *Marine Ecology Progress Series* 4, 9-21.

Craib, J.S. (1965) A sampler for taking short undisturbed marine cores. *Journal du Conseil Permanente International pour l'Exploration de la Mer, Copenhagen* 30, 34-39.

Dale, B. (1979) Collection, preparation, and identification of dinoflagellate resting cysts. In: Taylor, D.L. and Seliger, H.H. (eds) *Toxic Dinoflagellate Blooms*. Elsevier/North Holland, New York, pp. 443-452.

Eaton, J.W. and Moss, B. (1966) The estimation of numbers and pigment content in epipelic algal populations. *Limnology and Oceanography* 4, 584-595.

Ellegaard, M., Christensen, N.F. and Moestrup, Ø. (1994) Dinoflagellate cysts from recent Danish marine sediments. *European Journal of Phycology* 29, 183-194.

Fliermans, C.B. and Schmidt, E.L. (1977) Immunofluorescence for autoecological study of a unicellular blue-green alga. *Journal of Phycology* 13, 364-368.

Harris, A.D.S., Medlin, L.K., Lewis, J. and Jones, K.J. (1995) *Thalassiosira* species (*Bacillariophyceae*) from a Scottish sea-loch. *European Journal of Phycology* 30, 117-131.

Hasle, G.R. (1978) Some specific preparations: Diatoms. In: Sournia, A. (ed.) *Phytoplankton Manual*. UNESCO Press, Paris, pp. 136-142.

Hasle, G.R. and Fryxell, G.A. (1970) Diatoms: cleaning and mounting for light and electron microscopy. *Transactions of the American Microscopical Society* 89, 469-474.

Honjo, S. and Doherty, K. (1988) Large aperture time-series sediment traps: design objectives, construction, and application. *Deep Sea Research* 35, 133-139.

Honjo, S., Connell, J.F. and Sachs, P.L. (1980) Deep ocean sediment trap: design and function of Parflux Mark 2. *Deep-Sea Research* 27, 745-753.

Kanneworff, E. and Nicolaisen, W. (1973) The 'HAPS', a frame-supported bottom corer. *Ophelia* 10, 119-129.

Lewis, J., Dodge, J.D. and Tett, P. (1984) Cyst-theca relationships in some *Protoperidinium* species (*Peridiniales*) from Scottish sea lochs. *Journal of Micropalaeontology* 3, 25-34.

Lewis, J., Harris, A.D.S., Jones, K.J. and Edmonds, R.L. (1996) Long term survival of marine planktonic diatoms and dinoflagellates in stored sediment samples. *Journal of Plankton Research* (in press).

Matsuoka, K., Fukuyo, Y. and Anderson, D.M. (1989) Methods for modern dinoflagellate cyst studies. In: Okaichi, T., Anderson, D.M. and Nemoto, T. (eds) *Red Tides: Biology, Environmental Science and Toxicology*. Elsevier, New York, pp. 461-479.

Odate, T. and Maita, Y. (1990) Seasonal distribution and vertical flux of resting spores of *Chaetoceros* (*Bacillariophyceae*) species in the neritic water of Funka Bay, Japan. *Bulletin of the Faculty of Fisheries, Hokkaido University* 41, 1-7.

Ohnstad, F.R. and Jones, J.G. (1982) *The Jenkin Surface-Mud Sampler User Manual*. Occasional Publication 2, Freshwater Biological Association, Ambleside.

Palmer, J.D. and Round, F.E. (1965) Persistent, vertical-migration rhythms in benthic microflora. I. The effect of light and temperature on the rhythmic behaviour of *Euglena obtusa*. *Journal of the Marine Biological Association of the United Kingdom* 45, 567-582.

Pipe, A.E. and Cullimore, D.R. (1980) An implanted slide technique for examining the effects of the herbicide Diuron on soil algae. *Bulletin of Environmental Contamination and Toxicology* 24, 306-312.

Price, C.A., Reardon, E.M. and Guillard, R.R.L. (1978) Collection of dinoflagellates and other marine microalgae by centrifugation in density gradients of a modified silica sol. *Limnology and Oceanography* 23, 548-553.

Reid, P.C. and Boalch, G.T. (1987) A new method for the identification of dinoflagellate cysts. *Journal of Plankton Research* 9, 249-253.

Round, F.E. and Eaton, J.W. (1966) Persistent, vertical-migration rhythms in benthic microflora. III. The rhythm of epipelic algae in a freshwater pond. *Journal of Ecology* 54, 609-615.

Sharabi, N. El-Din and Pramer, D. (1973) A spectrofluorometeric method for studying algae in soil. *Bulletin of Ecological Research Communications* 17, 77-84.

Shubert, L.E. and Starks, T.L. (1979) Algal colonization on a reclaimed surface mined area in Western North Dakota. In: Wali, M.K. (ed.) *Ecology and Coal Resource Development*. Pergamon Press, New York, pp. 661-669.

Shubert, L.E. and Starks, T.L. (1985) Diagnostic aspects of algal ecology in disturbed lands. In: Tate, R.L. and Klein, D.A. (eds) *Soil Reclamation Processes*. Marcel Dekker, New York, pp. 83-106.

Starks, T.L., Shubert, L.E. and Trainor, F.R. (1981) Ecology of soil algae: a review. *Phycologia* 20, 65-80.

Throndsen, J. (1978) The dilution-culture method. In: Sournia, A. (ed.) *Phytoplankton Manual*. UNESCO Press, Paris, pp. 218-224.

Trainor, F.R. and Gladych, R. (1995) Survival of algae in a dessicated soil: a 35-year study. *Phycologia* 34, 191-192.

Uhlig, G. (1964) Eine einfache Methode zur Extraktion der vagilen, mesopsammalen Mikrofauna. *Helgoländer Wissenschaftliche Meeresuntersuchungen* 11, 178-185.

Wall, D. and Dale, B. (1968) Modern dinoflagellate cysts and evolution of the *Peridiniales*. *Micropaleontology* 14, 265-304.

Wiedeman, V.E., Walne, P.L. and Trainor, F.R. (1964) A new technique for obtaining axenic cultures of algae. *Canadian Journal of Botany* 42, 958-959.

Yamaguchi, M., Itakura, S., Imai, I. and Ishida, Y. (1995) A rapid and precise technique for enumeration of resting cysts of *Alexandrium* spp. (*Dinophyceae*) in natural sediments. *Phycologia* 34, 207-214.

Protozoa 7

JOHN F. DARBYSHIRE[1], **O. ROGER ANDERSON**[2] **AND ANDREW ROGERSON**[3]

[1]*Macaulay Land Use Research Institute, Craigiebuckler, Aberdeen AB9 2QJ, UK;* [2]*Biological Oceanography, Lamont-Doherty Earth Observatory of Columbia University, Palisades, NY 10964, USA;* [3]*University Marine Biological Station Millport, Isle of Cumbrae KA28 OEG, UK.*

Ecological Significance of Protozoa

A very diverse and large number of protozoan species, including flagellates, ciliates and amoebae, have been isolated from soils and sediments throughout the world (Fenchel, 1987; Anderson, 1988; Sleigh, 1989; Vickerman, 1992; Lee *et al.*, 1996). Frequently, these protists are dispersed by water currents and by wind as cysts. It has been estimated that there are around two cysts per cubic metrc of air (Sleigh, 1989). Photosynthetic species are significant in primary production. Non-photosynthetic heterotrophic species ingest microscopic prey or detritus and are important in the mineralization of organic matter with the release of nutrients and stimulatory metabolites for microorganisms and plants (Griffiths, 1994; Pussard *et al.*, 1994; Zwart *et al.*, 1994). Simple methods that could be used for estimating protozoan biodiversity in soils and sediments are briefly described. Whenever it is possible, several methods should be used in tandem and living organisms should be examined soon after sampling rather than relying solely on preserved specimens.

Methods for the Isolation of Protozoa from Soils and Sediments

Collection of Samples

The sampling strategy used should be designed to make the samples representative of the habitat as a whole and overcome local differences in the environment as much as possible. In practice, many sampling procedures are compromises between what is feasible in the time available and the requirements of other organisms under investigation in the same survey. Several small samples collected independently within the study area are preferable to one large sample, because they could be used in statistical evaluations of the results. Randomly positioned samples can be selected using pairs of random numbers obtained from tables (Fisher and Yates, 1963) for the co-ordinates of a grid system superimposed over the sampling area. Samples can then be collected at points where the two co-ordinates intersect. Alternatively, samples can be collected at regular intervals in one or two dimensions and then the spatial distribution of the protozoan species can be analysed by geostatistics (Rossi *et al.*, 1992). Sometimes, when the time available is limited, all the samples are bulked together to provide what is often termed a 'representative sample', although the effectiveness of this procedure is largely dependent on the number of samples and degrees of mixing used. It is usual to transport the samples in clean and sterile vessels within an insulated box (e.g. a cooled picnic container) so that the temperature of the samples remains unchanged.

Usually, augers or cork borers are used to collect soil samples. Large stones are removed by sieving the soil through coarse sieves. For samples near the water/sediment interface or within loose sediments, a large pipette attached to a rubber bulb is often used to collect samples. If the sediment is more compact and in shallow water, a spatula or spoon can be used to remove some sediment and transfer it to a collecting vessel with the minimum of disturbance. Simple Perspex corers can be used in shallow water sites and intertidal zones. In muddy sediments, the Perspex tube can be pushed into the substrate, the top sealed with a rubber bung and the whole tube removed from the mud. In sandy sediments, the surrounds of the inserted corer can be dug away and the bottom of the tube quickly sealed with a bung before the core is lifted. In the laboratory, both types of sediment core can be gradually extruded and cut into sections with a knife. Cork borers can be used in a similar manner to collect small samples of sediments or medical syringes, cut at their lower end, can be used as piston corers.

As the oxygenated zone in rich organic sediments is restricted to either the upper centimetre or even the first few millimetres, only the uppermost layer of such sediments should be collected when aerobic protozoa are being investigated. Anaerobic protozoa can usually be collected from sediment below 1 cm and deeper. In sandy sediments with larger individual aggregates, the aerobic zone is much more extensive, often extending to more than 15 cm depth. A pole corer with an automatic closing top seal can be used for collecting samples from shallow sublittoral soft sediments (Frithsen *et al.*, 1983; Ali, 1984). In deeper sublittoral sediments, a remote corer device should be used. The most reliable type is the bulky 'Craib' corer, which needs to be lowered from a ship's winch (Craib, 1965). Several modifications of the 'Craib' corer have been developed and these can be obtained from Messrs. Bowers and Connelly, c/o Dunstaffnage

Marine Laboratory, Oban, UK. Multicore versions are available and can provide quantitative samples from the deep sea.

Decanting Interstitial Water

The simplest method to extract protozoa from wet sediment or soil is to decant off the aqueous phase collected with the sample. The solid material can either be held to one side of a tilted Petri dish to allow the water to drain off or larger volumes can be processed by straining through a nylon sieve with a pore size in the range 5-30 μm. After draining, additional filtered water can be added for further extraction from the same sediment. Interstitial protozoa are sampled by this method, particularly free-swimming ciliates and flagellates.

Non-flooded Petri Dish Method

Foissner (1992) described a similar method that enabled him to isloate many new species of soil protozoa. His protocol was as follows:

- Place 10-15 g of fresh or air-dried soil in a Petri dish of 10-50 cm diameter.
- Saturate the soil, but do not flood the sample with distilled water. The water should be added until 5-20 ml surplus water will drain off when the Petri dish is tilted to 45° from the horizontal and the soil is gently pressed with a finger. Complete saturation may require up to 12 h.
- Cover the Petri dish with a loose-fitting lid to allow some gas exchange.
- Examine small aliquots of the liquid from the Petri dish on 2, 6, 12, 20 and 30 days after the soil was saturated. Further inspections usually provide only a few new species.

Migration on Agar

Many small soil and sediment protozoa, especially amoebae and flagellates, can be isolated on agar surfaces. These protozoa consume the attendant bacteria, reproduce and migrate in the thin film of liquid on the agar surface away from the site of inoculation. A variety of different agar-based media can be used. For soil organisms, non-nutrient agar with a streak of bacteria (e.g. *Escherichia coli*) is a good choice. For sediment protozoa, cerophyl agar, malt/yeast extract agar and soil extract agar should be considered. In all cases, a small inoculum of soil or sediment (enough to cover an area with a 0.5 cm diameter) is placed on the agar surface. If the agar is enriched with a bacterial streak, the soil/sediment should be placed at one end of the streak. The plates are sealed with a strip of plastic cling wrap to prevent moisture loss, inverted and incubated for two to three weeks. Subsequently, protozoa can be washed off the agar surface with sterile water from the collection site and examined under a microscope.

Colonization of Glass Slides or Chambers

A convenient sample of sessile or thigmotactic species can be obtained by placing acid-washed, or agar-coated, glass microscope slides on the surface of sediments or soils. Alternatively, slides can be inserted vertically to isolate protozoa from different depths. After several hours or days, depending on the protozoan diversity and abundance, the slides can be carefully withdrawn and examined for adhering species. It is recommended that the slides are quickly protected from drying out after removal from the substrate either by covering the slides with large cover slips and viewing as soon as possible from above with a conventional microscope or by submerging them in sterile water in a Petri dish and examining with an inverted microscope. Some of the adhering protists may only survive on the slides for a short time after removal from the soil or sediment. This method is particularly suitable for observing species that cannot be cultured in the laboratory. Large reticulate foraminifera, naked amoebae, thigmotactic flagellates with trailing flagella and sessile ciliates can be sampled by this method. If a numerical estimate of diversity is required, then the number of different species in several randomly selected fields of view should be counted. It is important to use a microscope objective that allows one to identify species and still provides a representative field of view. Usually, counts from about 20 random fields are sufficient to obtain a representative sample.

Flattened microcapillaries can be used in a similar fashion to glass slides. Such precision-made capillaries, with an inner thickness of 0.05-0.40 mm, have been used by Russian soil microbiologists for many years (Perfilev and Gabe, 1969) and can now be purchased ready-made elsewhere (Camlab, Cambridge, UK). These capillaries are less liable to dry out than glass slides, can be used with nutrient media and can help to start clonal cell cultures. To reduce the chances of breakage in soil or sediment, the glass capillaries are usually attached to microscope slides temporarily. Capillary chambers of other designs can easily be made by attaching long cover slips to slides. Glass spacers made from fragments of cover glasses can be used to raise the cover slip above the surface of the slide and the entire assembly secured by a suitable glue, for example., petroleum jelly or 'Araldite' (an epoxy resin glue).

Substrate Dilution with or without Nutrient Enrichment

In the natural state, many protozoan species occur in low numbers and may be obscured by individual soil or sediment particles. Dilution of the soil or sediment, often followed by nutrient amendment of the dilution series, frequently encourages some of these protozoa to multiply and can result in larger biodiversity estimates than those based entirely on the methods described earlier.

For soils, a 10 g fresh weight sample is usually suspended in 50 ml of sterile liquid soil extract and shaken for 5 min on an orbital shaker at 60 revs min^{-1} (Darbyshire, 1973). After this time, wide-mouthed pipettes are used to pipette aliquots (0.05 ml) into the first row of 12 wells of sterile rigid styrene plates (Falcon Plastics Ltd., supplied by Becton Dickinson, Wembley, UK), each plate containing a total of 96 wells. Previously, all the other wells in these styrene plates are filled with 25 μl of liquid soil extract using either microdiluters (Dynatech Laboratories Ltd., Billinghurst, UK) or multipipette dispensers. Liquid soil extract is prepared from air-dried samples of the soil under investigation. Starting from the row with the original dilution, twofold dilution series of the soil samples are then prepared by transferring 25 μl aliquots from the wells of one row to the corresponding wells in the next row using either microdiluters or multipipettes. A comparison between this method and that of Singh (1946, 1955), in which 0.5% (w/v) NaCl solution is used as the diluent and every well is amended with a thick bacterial suspension of *Klebsiella aerogenes*, showed that more protozoan species were obtained with this liquid soil extract modification (Darbyshire, 1973). The plates are incubated for four weeks at 20°C inside containers with small beakers of water to reduce evaporation from the soil dilutions. At 4, 7, 10, 14, 21 and 28 days after inoculation, each soil dilution is examined with an inverted microscope using long-focus objectives with total magnifications up to × 800 and the presence of different protozoan species noted. This method is often referred to as the most probable number (MPN) method. Several computer programs can be used for these calculations (Hurley and Roscoe, 1983).

For sediments, another less elaborate enrichment cultivation method has been found to provide comparable estimates of bactivorous protozoa to the MPN method just described. In this method, a known volume or weight of sediment is suspended in a measured volume of membrane-filtered water from the sampling site. Aliquots are inoculated into previously prepared culture dishes, as described below. It is often necessary to try several different dilutions to determine which dilution provides the optimum protozoan yield. Ideally, populations should develop from single-cell inocula (each cell deposited in one well of a multi-well culture plate). At least two sterile rigid styrene plates, each containing 24 wells (Falcon Plastics Ltd., supplied by Becton Dickinson, Wembley, UK), are used for each sediment sample. Each well is amended with a small cube (1 mm^3) of malt/yeast extract agar or cerophyl agar to encourage growth of bacterial prey for protozoa as well as 2 ml of membrane-filtered water from the sampling site and an aliquot of the diluted sediment (often 0.1 ml). The diluted sediment suspension is thoroughly mixed before the aliquots are added to each well. The culture plates are covered by lids to reduce evaporation and incubated for 12-14 days. After this period, each well is examined with an inverted microscope with phase-contrast objectives of at least × 40 magnification. A list of protozoan species in each well is compiled and the density of each species per unit volume or weight is calculated from the formula:

$$D = \frac{[N \times V/I \times W]}{S}$$

where D = density of protozoan species,
N = number of wells containing the species of interest,
V = volume of sediment suspension used to prepare inoculum,
I = size of aliquot of suspension added to each well,
W = total number of wells in all culture plates used and
S = quantity of sediment (in ml or g dry weight) used to prepare sediment suspension.

For the isolation of algaliferous protozoa, small quantities of heat-killed *Chlorella* sp. can be used in the wells to promote growth. This heat treatment prevents the *Chlorella* from swamping the growth of other microbes in the wells.

Seawater Ice Method for Marine Sediments

This method was originally described by Uhlig (1964) and is appropriate for the isolation of motile, interstitial protozoa. The bottom end of a glass or plastic tube with the same diameter as a sediment core is securely covered with nylon mesh. Some of the sediment sample is placed inside the tube and covered with a layer of cotton wool. The sediment should occupy about 20% of the volume of the tube with a slightly smaller volume for the cotton wool. The remainder of the tube is filled with seawater ice. The tube is suspended above a crucible or similar collecting vessel filled with membrane-filtered seawater, so that the lower end of the tube is just in contact with the sea water. As the ice slowly melts, the salinity of the sediment gradually increases and the protozoa migrate out of the sediment into the underlying reservoir of seawater. This seawater reservoir should be examined for protozoa and regularly replaced by fresh membrane-filtered seawater. The temperature gradient created is not believed to be a significant factor. If required, the protozoa can be preserved for later analysis. Additional details can be found in Carey (1992). Small (1992) described a modification where seawater ice was replaced by filtered water from the sampling site and the interstitial protists are gradually eluted or migrate from the sediment. Although it has not been investigated, it is possible that some soil protozoa can be extracted from soil cores by this method.

Sieving Larger Protozoa

Shelled protozoa (e.g. foraminifera and testate amoebae) can be isolated from sediments and soils by sieving suspensions of these samples through fine membranes (Coûteaux, 1967; Thiel, 1983; Gooday, 1986). For fine-grained sediments, a core sample of approximately 10-30 cm^2 and 5 cm depth is adequate. Such a sediment sample, either fresh or preserved is 'wet'-sieved through a series of filters of progressively narrower pores down to approximately 63-45 μm diameter using membrane-filtered water from the collection site. Lugol's iodine or rose bengal is often used to stain the protozoan cytoplasm to distinguish living from dead cells, but a fluorochrome (acetoxymethyl esters of biscarboxyethylcarboxyfluorescein) that has recently been described as a good stain for active foraminifera (Bernhard *et al.*, 1995) could be used instead. Larger species can be isolated by probing the diluted sediments with fine needles. Testate amoebae can be isolated from soil suspensions after fixation with Bouin by filtering well-mixed soil suspensions through fine membranes (Millipore Ltd., 2 μm diameter pores). The membranes are mounted in immersion oil on glass microscope slides and observed with a microscope (Coûteaux, 1967).

Density Gradient Centrifugation

Schwinghamer (1981) separated live foraminifera from marine sediments by centrifugation with a mixture of 'Percoll' silica gel and sorbitol. Alongi (1986) separated a wide variety of ciliates, flagellates and amoebae besides foraminifera with the same technique from small samples of marine sediments. Alongi extruded his sediment samples (from 2 cm depth with 0.64 cm^2 surface area) into 30 ml centrifuge tubes containing 5 ml aliquots of a 'Percoll'/sorbitol mixture, then he vortexed the samples for 1-2 min, allowed them to stand for 1 h and finally centrifuged the samples at 490 *g* for 20 min. The supernatants were decanted and examined for protozoa. The extraction procedure was repeated three times with each sample. In a later publication, he suggested that a range of concentrations of 'Percoll' and sorbitol should be tested to determine the optimal mixture for each sediment (Alongi, 1991). Griffiths and Ritz (1988) similarly used density gradient centrifugation (3010 *g* for 2 h) with a mixture of 'Percoll' and Sørensen's phosphate buffer (pH 7.0) to separate a range of small active protozoa from a mineral soil.

Cultivation of Microaerophilic Species

Recently, there has been interest in anaerobic or microaerophilic protozoa normally found living deeper in sediments below the oxic/anoxic boundary (Fenchel and Finlay, 1995). As detectable levels of oxygen are toxic for these species, they must be isolated and cultivated under low oxygen tensions. In practice, it is very difficult to eliminate all traces of oxygen from cultures. The simplest method of removing most of the oxygen is to exploit the activities of aerobic bacterial contaminants. A plug of sediment from the anoxic zone should be placed at the bottom of a test tube filled with filtered water from the same site. The addition of a few rice grains will encourage the growth of aerobic bacteria and remove the last traces of detectable oxygen. At this stage, the culture conditions will be appropriate for the growth of anaerobic protozoa. If larger volumes of sediment are involved, then the samples should be added to bottles containing boiled and degassed filtered water. These bottles should be sealed with a crimp capping system and further deoxygenated by bubbling nitrogen through the sediment.

Routine for Handling Samples on Return to the Laboratory

After some interstitial water from each sample is removed and quickly surveyed under a microscope, the remainder of the samples should be divided equally by weight or volume between the isolation methods chosen from those described. The density gradient centrifugation method will provide results most rapidly; the others require several weeks' incubation and observation. The best temperatures for incubation are between 15 and 20°C. Identifications based on several living and stained specimens are to be preferred.

Reagents and Media

Soil Extract (stock)
Soil is passed through a 200 mm diameter sieve (4.76 mm mesh, Endecott Filters Ltd., London, UK). One kg of the sieved soil is mixed with 1 litre of tap water and autoclaved (121°C for 20 min). The supernatant of the soil suspension is decanted, filtered (size No. 2 filter, British Berkefield Ltd., Tonbridge, UK) and finally autoclaved.

Soil Extract Medium (for marine protozoa; Page, 1983)
Prepare soil extract stock as detailed above and two stock solutions (a) 20 g of $NaNO_3$ in 100 ml distilled water and (b) 1.18 g Na_2HPO_4 in 100 ml distilled water. Combine 950 ml filtered seawater with 50 ml of soil extract and 1 ml of each stock solution (a and b) and autoclave. This liquid medium (1 litre) can be solidified with 15 g non-nutrient agar l^{-1}.

Soil Extract Medium (for freshwater/soil protozoa; Page, 1988)
Prepare soil extract stock as detailed above and combine 10 ml soil extract with 84 ml distilled water and 2 ml of three stock solutions of salts: (1) K_2HPO_4, 0.1% w/v, (2) $MgSO_4.7H_2O$, 0.1% w/v, (3) KNO_3, 1.0% w/v. This liquid (100 ml) can be solidified with 1.5 g non-nutrient agar l^{-1}.

Malt/Yeast Extract Agar (Page 1983, 1988)
Add 15 g non-nutrient agar to 1 litre filtered seawater or spring water containing 0.1 g malt extract and 0.1 g yeast extract. Dissolve the powders and autoclave.

Cerophyl Agar (Page, 1983, 1988)
Filter 1 litre of water from the collection site and add 1 g of cereal leaves (e.g. Cerophyl or Sigma cereal leaves). Boil for 5 min and filter through a glass-fibre filter and restore volume to 1 litre. Solidify with 15 g non-nutrient agar l^{-1}.

Lugol's Iodine
This is a saturated solution of iodine in KI. To prepare, dissolve 6 g potassium iodide in 20 ml distilled water. Then add iodine (4 g) to the KI solution and make up the volume to 100 ml with distilled water.

Bouin Fixative 1
Saturated aqueous solution of picric acid (60 ml), 40% formalin (20 ml) and glacial acetic acid (4 ml).

Bouin Fixative 2
Saturated aqueous solution of picric acid (75 ml), 40% formalin (25 ml) and trichloroacetic acid (2 g).

Rose Bengal Stain (Boltovskoy and Wright, 1976)
Dissolve 1 g of rose bengal in distilled water. Alternatively, a solution of 1 g of rose bengal, 5 ml phenol and 100 ml of distilled water is very effective. The rose bengal solution is added to the sample and allowed to sit for 10-20 min, after which time the excess colouring agent is washed off with distilled water.

Bibliography

Ali, A. (1984) A simple and efficient sediment corer for shallow lakes. *Journal of Environmental Quality* 13, 63-66. [Description of a pole corer designed for use in shallow lakes. The tight closure of the coring tube by a stopper also enables sampling of the flocculent surface layer.]

Alongi, D.M. (1986) Quantitative estimates of benthic protozoa in tropical marine systems using silica gel: a comparison of methods. *Estuarine, Coastal and Shelf Science* 23, 443-450. [Comparison of silica gel method with 7 other extraction techniques for marine protozoa from sediments.]

Alongi, D.M. (1991) Flagellates of benthic communities: characteristics and methods of study. In: Patterson, D.J. and Larsen, J. (eds) *The Biology of Free-Living Heterotrophic Flagellates*. Clarendon Press, Oxford, pp. 57-73. [Recent discussion of the use of 'Percoll'/sorbitol mixtures in density gradients.]

Anderson, O.R. (1988) *Comparative Protozoology*. Springer-Verlag, Berlin. [Advanced text including the major principles of ecology, physiology and life history of major groups of protozoa in a comparative context.]

Bernhard, J.M., Newkirk, S.G. and Bowser, S.S. (1995) Towards a non-terminal viability assay for foraminiferan protists. *Journal of Eukaryotic Microbiology* 42, 357-367. [Recent description of the use of different fluorochromes for distinguishing active *Foraminifera*.]

Boltovskoy, E. and Wright, R. (1976) *Recent Foraminifera.* Junk, The Hague. [Comparative survey of major topics in foraminiferan biology and paleontology with methods for analysis of sediment samples.]

Carey, P.G. (1992) *Marine Interstitial Ciliates, An Illustrated Key.* Chapman and Hall, London. [Comprehensive identification guide for interstitial ciliates around the UK, with an introductory section on practical methods for maintenance, extraction and observation.]

Coûteaux, M.M. (1967) Une technique d'observation des Thécamoebiens du sol pour l'estimation de leur densité absolue. *Revue d'Écologie et de Biologie du Sol* 4, 593-596. [Original description of Coûteaux membrane technique.]

Craib, J.S. (1965) A sampler for taking short undisturbed marine cores. *Journal du Conseil Permanent International pour l'Exploration de la Mer* 30, 34-39. [This paper describes the Craib corer, a design still preferred for sampling fine sediment in deep water. The device has a ball valve which seals the bottom of the core tube when it is withdrawn from the sediment.]

Darbyshire, J.F. (1973) The estimation of soil protozoan populations. In: Board, R.G. and Lovelock, D.W. (eds) *Sampling - Microbiological Monitoring of Environments.* Academic Press, London, pp. 175-188. [Review of MPN technique for soil protozoa.]

Fenchel, T. (1987) *Ecology of Protozoa.* Science Tech Publishers, Madison. [Comprehensive survey of the ecology of protozoa including physiological ecology and energetics of protozoan species and communities.]

Fenchel, T. and Finlay, B.J. (1995) *Ecology and Evolution in Anoxic Worlds.* Oxford University Press, Oxford. [A book on the natural history of anoxic environments and their microbial inhabitants with emphasis on their metabolic pathways.]

Fisher, R.A. and Yates, F. (1963) *Statistical Tables for Biological, Agricultural and Medical Research,* 6th edn. Oliver and Boyd, Edinburgh. [Standard set of statistical tables.]

Foissner, W. (1992) Estimating the species richness of soil protozoa using the 'non-flooded petri dish method'. In: Lee, J.J. and Soldo, A.T. (eds) *Protocols in Protozoology*, Allen Press, Lawrence, pp. B.10.1-B.10.2. [Description of non-flooded Petri dish method.]

Frithsen, J.B., Rudnick, D.T and Elmgren, R. (1983) A new, flow-through corer for quantitative sampling of surface sediments. *Hydrobiologia* 99, 75-79. [Description of a pole corer for sampling shallow sublittoral soft sediments. This design is simple and has a reliable closing lid.]

Gooday, A.J. (1986) Meiofaunal foraminiferans from the bathyl Porcupine Seabight (north-east Atlantic): size structure, standing stock, taxonomic composition, species diversity and vertical distribution in the sediment.

Deep-Sea Research 33, 1345-1373. [Survey of benthic foraminifera across a wide size range provides information on standing crops and significance of smaller foraminifera in fine sieve fractions from benthic habitats.]

Griffiths, B.S. (1994) Soil nutrient flow. In: Darbyshire, J.F. (ed.) *Soil Protozoa*. CAB INTERNATIONAL, Wallingford, pp. 65-91. [Recent review of subject.]

Griffiths, B.S. and Ritz, K. (1988) A technique to extract, enumerate and measure protozoa from mineral soils. *Soil Biology and Biochemistry* 20, 163-173. [Description of isolation method for soil *Protozoa* using density gradients of 'Percoll' and Sørensen's phosphate buffer.]

Hurley, M.A. and Roscoe, M.E. (1983) Automated statistical analysis of microbial enumeration by dilution series. *Journal of Applied Bacteriology* 55, 159-164. [Contains computer program for MPN counts.]

Lee, J.J., Leedale, G.F., Patterson, D.J. and Bradbury, P. (eds) (1996) *Illustrated Guide to Protozoa*, 2nd edn. Allen Press, Lawrence. [Recent discussion of classification and diversity of *Protista*.]

Page, F.C. (1983) *Marine Gymnamoebae*. Institute of Terrestrial Ecology, Cambridge. [Guide to the identification of marine naked amoebae with introductory sections on methods for cultivation, observation and study.]

Page, F.C. (1988) *A New Key to Freshwater and Soil Gymnamoebae*. Freshwater Biological Association, Windermere. [Guide to the identification of freshwater and soil naked amoebae with notes on media, culture and methods of observation.]

Perfilev, B.V. and Gabe, D.R. (1969) *Capillary Methods of Investigating Microorganisms*. (English translation by J.M. Shewan.) Oliver and Boyd, Edinburgh. [Detailed description of the construction and use of glass capillaries.]

Pussard, M., Alabouvette, C. and Levrat, P. (1994) Protozoan interactions with the soil microflora and possibilities for biocontrol of plant pathogens. In: Darbyshire, J.F. (ed.) *Soil Protozoa*. CAB INTERNATIONAL, Wallingford, pp. 123-146. [Recent review of subject.]

Rossi, R.E., Mulla, D.J., Journel, A.G. and Franz, E.H. (1992) Geostatistical tools for modelling and interpreting ecological spatial dependence. *Ecological Monographs* 62, 277-314. [Review of the use of geostatistics in ecology.]

Schwinghamer, P. (1981) Extraction of living meiofauna from marine sediments by centrifugation in a silica sol-sorbitol mixture. *Canadian Journal of Fisheries and Aquatic Sciences* 38, 476-478. [Early description of density gradients of 'Percoll'/sorbitol mixtures for isolating live foraminfera.]

Singh, B.N. (1946) A method of estimating the numbers of soil protozoa, especially amoebae, based on their differential feeding on bacteria. *Annals of Applied Biology* 33, 112-119. [Original description of Singh's ring method.]

Singh, B.N. (1955) Culturing soil protozoa and estimating their numbers in soil. In: Kevan, D.K.McE. (ed.) *Soil Zoology*. Butterworth, London, pp. 403-411. [Further description of Singh ring method.]

Sleigh, M. (1989) *Protozoa and Other Protists*. Edward Arnold, London. [Excellent introduction to major concepts of protistology including protozoa and smaller algae.]

Small, E.B. (1992) A simple method for obtaining concentrated populations of protists from sediments. In: Lee, J.J. and Soldo, A.T. (eds) *Protocols in Protozoology*, Allen Press, Lawrence. pp. B.3.1-B.3.4. [Recent modification of Uhlig seawater ice method.]

Thiel, H. (1983) Meiobenthos and nanobenthos of the deep-sea. In: Rowe, T. (ed.) *The Sea,* Vol. 8. John Wiley, New York, pp. 167-230. [Methods for collection, separation, observation and enumeration of a wide range of benthic-dwelling biota with a concise introduction to the major taxonomic groups.]

Uhlig, G. (1964) Eine einfache Methode zur Extraktion der Vagilen, mesopsammalen Mikrofauna. *Helgolander Wissenschaftliche Meeresuntersuchungen* 11, 178-185. [Original description of seawater ice method.]

Vickerman, K. (1992) The diversity and ecological significance of protozoa. *Biodiversity and Conservation* 1, 334-341. [Recent review of the subject.]

Zwart, K.B., Kuikman, P.J. and Van Veen, J.A. (1994) Rhizosphere protozoa: their significance in nutrient dynamics. In: Darbyshire, J.F. (ed.) *Soil Protozoa*. CAB INTERNATIONAL, Wallingford, pp. 93-121. [Recent review of the subject.]

Dictyostelids and Myxomycetes 8

STEVEN L. STEPHENSON[1] AND JAMES C. CAVENDER[2]

[1]*Department of Biology, Fairmont State College, Fairmont, WV 26554-2491, USA;* [2]*Department of Environmental and Plant Biology, Ohio University, Athens, OH 45701-2979, USA.*

Dictyostelids and Myxomycetes in Terrestrial Ecosystems

A major portion of the net annual primary production in forests and other terrestrial ecosystems becomes directly or indirectly available to the decomposers of the detritus food chain. These decomposers (bacteria and fungi) are, in turn, an important food resource for various phagotrophic invertebrates and protozoans. Bacteria, for example, are preyed upon by bacterivores (e.g. protozoa and nematodes) as well as some detritivores (e.g. earthworms). Naked amoebae, which can make up 95% of the protozoan population in some soils, are the most important group of protozoa with respect to bacterial consumption. In addition to their direct influence on the structure of soil microbial communities, these amoebae play a key role in nutrient cycling. Mineralization is stimulated and decomposition enhanced by the amoebae releasing nutrients tied up in the microbial biomass. For example, amoebae are known to release ammonia to plant roots when feeding on bacteria and can produce increases in dry weight and nitrogen content. It is not known what percentage of the total population of soil amoebae is made up of the amoeboid stages of dictyostelids (cellular slime moulds) and myxomycetes (plasmodial slime moulds), but, judging from observations of amoeboid colonies on our isolation plates and the data available from a number of recent studies, this percentage is significant. Interestingly, dictyostelids are relatively more numerous than myxomycetes in forest soils, whereas myxomycetes are usually more abundant than dictyostelids in grassland and agricultural soils.

Collection of Soil Samples

Several different methods have been used to collect soil samples for the isolation of dictyostelids and myxomycetes. These include scraping the soil surface with the mouth of the collecting container (e.g. a small plastic vial), scraping surface soil into a 'Whirl-Pac' or other plastic bag with a wooden tongue depressor (which is discarded after being used for a particular sample), removing a bulk sample of soil collected from a depth of several centimetres and placing this in a suitable container, and collecting soil cores with the use of some type of hollow cylinder. Whatever the method used, an effort should be made to ensure that the sample is representative of the particular site or study area from which it is collected.

If statistical analysis of the resulting data is to be undertaken, this must be considered when designing the sampling scheme to be used. For example, Feest and Madelin (1985b) and Feest (1986) used a set of samples collected from a single 1 m^2 quadrat to characterize the myxomycetes present in soil at a particular study site. The quadrat was divided internally into one hundred 10 dm squares and 60 cylindrical cores of 1 cm^2 cross-sectional area were removed from the upper 4 cm of the soil profile in squares chosen with the use of a random number table. The 60 cores were then bulked together to form one sample.

Stephenson and Landolt (1995) used a similar sampling method to examine the vertical distribution of myxomycetes and dictyostelids in forest soils at a study site in Virginia, USA. Cores were extracted using an aluminium cork borer with an inside diameter of 2.4 cm. To obtain a core, the cork borer was pressed firmly against the ground and enough force applied to cause the borer to penetrate the soil to a depth of approximately 4-5 cm. After the borer had been removed from the soil, the intact core was extruded and samples from the portions of the core representing the three layers being considered were removed and transferred to individual sterile plastic bags. Three to seven cores were taken to obtain what was arbitrarily considered to represent an adequate sample (usually ≥ 2 g fresh weight).

A small (no more than 3-5 g is needed) subsample of each soil sample provides sufficient material for the isolation and enumeration techniques used for dictyostelids and myxomycetes. However, for some forest soils (e.g. those of dry tropical forests) a larger composite sample may be necessary to enumerate all of the species present. The soil samples can be held in a refrigerator for several days without significant losses of taxa and/or numbers, so processing of samples need not be done immediately. When samples are processed, aliquots obtained from the same subsample can be used for both dictyostelids and myxomycetes.

Isolation Techniques

Dictyostelids

There are 60-70 known species of dictyostelids, the primary habitat of which is forest soil, although most soils contain some of them. The highest numbers occur at the soil surface and abundance decreases rapidly with increasing depth below the humus layer. On a global scale, a linear decrease in species richness of dictyostelids occurs with an increase in latitude and diversity has been found to be greatest in tropical and subtropical forests of Central and South America. Sampling methods must take account of their life cycle, food sources and moisture requirements.

Life cycle

The asexual life cycle of dictyostelids involves a phagocytic amoeboid feeding phase, during which the independent cells repel one another, an aggregation phase, during which the cells attract one another, and a culmination phase, during which the cells differentiate into supportive stalk cells formed into a sorocarp, or spores which enable the species to be disseminated (Raper, 1984; Hagiwara, 1989). Their generation time is about 3-4 h at 22°C, during which each cell will consume about 1000 bacteria; spore to spore time is around 72 h. Alternatively, sexual fusion may occur during pre-aggregation, resulting in the formation of multicellular resistant structures called macrocysts instead of sorocarp formation. Aggregation is brought about by production of chemotactic agents (acrasins), which are known to be simple molecules such as cAMP (in *Dictyostelium discoideum*), a dipeptide (in *Polysphondylium violaceum*) or a pterin (in *Dictyostelium lacteum*).

Food sources

Dictyostelium discoideum, one of the most thoroughly investigated species, can feed on a wide variety of saprophytic and pathogenic bacteria from such diverse families as the *Enterobacteriaceae, Pseudomonadaceae, Bacillaceae, Micrococcaceae, Nitrobacteriaceae* and *Spirillaceae*. Several taxa pathogenic to plants and animals, such as *Salmonella* and *Erwinia,* are also rapidly consumed by *D. discoideum. Dictyostelium mucoroides* may be cultured on many common and rare soil bacteria, although the quality of the bacterial food supply affects growth and sorocarp formation. In a laboratory study, it was found that each dictyostelid species preferentially phagocytoses certain bacterial strains, outcompeting other species when grown with that particular strain (Horn, 1971). The discontinuous distribution of bacterial colonies in nature limits direct competition between species and results in a patchy distribution of species in small areas of soil.

In forest soils, fluctuations in the density of dictyostelids correlate with fluctuations in the density of bacteria, particularly the numbers of aerobic bacteria, and dictyostelids therefore assume the role of indicators of bacterial populations. They have also been used to monitor the recovery of soil microbial populations after disturbance by slash-and-burn agriculture in a tropical forest environment (Cavender, 1993). Under the conditions of the study, recovery of dictyostelid populations was estimated to take about 14 years. An estimate for recovery from agricultural disturbance in a temperate forest, based on a similar study carried out in New York State, was at least 75 years.

Moisture requirements

Moderate moisture levels are required for dictyostelids to spread in soil. A minimum level of 15% soil moisture is necessary and colonization rates improve as moisture levels increase from 15 to 30%. Distributional studies have shown that intermediate moisture regimes are optimal for dictyostelids and the soil population tends to be most active in the autumn in temperate regions. In an oak-beech forest in Delaware, USA, active forms (myxamoebae) made up 48-51% of the total population in autumn, with 24% active forms in spring, and only 10-12% active forms found in summer and winter.

Cellular slime moulds are easily isolated from soil samples collected in nature (Cavender and Raper, 1965). For example, 10 g of forest humus in Ohio, USA, will yield four to eight species of dictyostelids identifiable in isolation plates. There are five important factors to be considered in the isolation of dictyostelids from soil (Cavender, 1990):

- use of a low-nutrient medium lacking inhibitors; a hay (not to be confused with agricultural hay) infusion medium made from aged mature lawn grass (*Poa* spp.) has been shown to work well,
- some buffering capacity; small amounts of phosphate buffers (0.2%) are used, giving a pH of *ca.* 6.2,
- low aqueous soil dilution (< 1:50) done in two stages; this removes coarse particles and most inhibiting factors,
- provision of a pre-grown bacterial food source; *Escherichia coli* strain Blr has been found to be excellent, inhibiting the development of most soil fungi and bacteria, and
- use of undried, unfrozen surface soil and leaf mould; freezing the soil kills trophic cells and provides a means of determining percentages of active and encysted forms.

The dictyostelids with smaller fruiting structures are the most difficult to isolate since they are the most sensitive to inhibiting factors. Dictyostelids are easily cultured in two-membered culture using a pre-grown bacterial streak (*Escherichia coli*) on a 1.5% (w/v) non-nutrient agar.

The technique utilized for isolating dictyostelids can also be used to obtain quantitative data on population sizes, since the colonies appear as cleared areas on the isolation plates and these can be counted. Collection of 5-10 samples from a particular area will produce all or most of the species present and allow ecological comparisons with other areas. More collecting is often required to produce the rarest species. The greatest limitations are in the preparation of the hay infusion medium, which must have a weak nutrient composition or inhibiting fungi and soil bacteria may overgrow the culture. Hay is a poor term for describing the main ingredient used in preparing the medium, since the best substrate has been found to be mature aged lawn grass, mostly *Poa* spp. Defined media have not been found to work as well.

Numbers of dictyostelids found in soil usually range from 100-200 per gram, to several hundred per gram, or up to one or two thousand per gram. Occasionally, soil samples are collected which are devoid of dictyostelids. These organisms can be spread in the soil by earthworms and arthropods, salamanders, small mammals and birds. Their numbers are known to fluctuate within a two-week period and probably on a daily basis as has been shown for other protozoa. Soil predators such as nematodes rapidly consume dictyostelid amoebae and spores, therefore necessitating the constant reintroduction of propagules by animals and birds.

Myxomycetes

The myxomycete life cycle involves two very different trophic stages, one consisting of uninucleate amoebae, with or without flagella, and the other consisting of a distinctive multinucleate structure, the plasmodium (Martin *et al.*, 1983). Under favourable conditions, the plasmodium gives rise to one or more fruiting bodies containing spores. The spores complete the life cycle by germinating to produce the uninucleate amoeboflagellate cells. The amoeboid stage feeds, grows, and multiplies by fission to produce large cell populations which produce resistant cysts under adverse conditions or the coenocytic plasmodium under continued favourable conditions. Until quite recently, plasmodial formation in all myxomycetes was thought to be sexual and heterothallic, with haploid uninucleate amoeboflagellate cells of compatible mating types acting as gametes and fusing to produce a diploid plasmodium. The fruiting bodies formed from the latter then produced spores by meiosis. However, some myxomycetes are now known to be apomictic, with the same level of ploidy existing throughout the life cycle. Both systems have been found to exist in some species. A single morphospecies may be made up of heterothallic, as well as non-heterothallic (i.e. apomictic) individuals, and it is even possible for a conversion from one system to the other to occur.

Both trophic stages in the myxomycete life cycle characteristically occur in microhabitats where bacteria are plentiful, and there seems little doubt that bacteria constitute an important (if not the major) source of food for these organisms. Myxomycete plasmodia are capable of feeding upon a wide variety of bacteria, although not all types are equally suitable for growth. Little appears to be known about the range of bacteria on which the amoeboflagellate cells feed. Although several types of bacteria (e.g. *Escherichia coli* and *Klebsiella aerogenes*) have been used as a food source in laboratory studies, some types do appear to be more favourable than others.

There are approximately 700 recognized species of myxomycetes, many of which are cosmopolitan and common inhabitants of terrestrial (particularly forest) ecosystems throughout the world. Most ecological studies have focused on those species characteristically associated with decaying wood or bark (i.e. coarse woody debris) on ground sites (Alexopolous, 1963; Stephenson, 1989). The species found in such microhabitats often occur in great profusion, typically produce fruiting bodies of sufficient size to be easily detected in the field, and certainly represent the best-known myxomycetes both taxonomically and aesthetically (Martin and Alexopoulos, 1969). Myxomycetes have been recorded on, or isolated from, soil on a number of occasions (Thom and Raper, 1930; Warcup, 1950; Indira, 1968), but only recently have the myxomycetes associated with the soil microhabitat been studied to any real extent.

Feest and Madelin (1985a) described a method for the selective isolation of myxomycetes from soils that uses culture plates prepared with half-strength corn meal agar (CMA/2) and inoculated with a suspension of washed cells of *Saccharomyces cerevisiae*. The abundance of myxomycetes, measured in terms of plasmodium-forming units (PFUs), was determined by a most probable number procedure. Distributional studies using this method showed that myxomycetes are rather common organisms in most soils, with the highest numbers generally being recorded from grassland and agricultural soils (Feest and Madelin, 1985a,b; 1988a,b). In these and other studies, myxomycetes have been shown to have a clumped distribution in soils and to exhibit the same decline in numbers with increasing soil depth typical of dictyostelids. Species of *Didymium* appear to be the most widespread and abundant myxomycetes present in soils in temperate regions of the world, where the majority of studies have been carried out (Feest and Madelin, 1988a), although tropical and subtropical soils are more likely to yield other taxa (Feest, 1987).

Kerr (1994) used a slightly modified version of the technique described by Feest and Madelin (1985a) in a survey of myxomycetes in soils of the northern United States. In this study, samples were collected from the top 4 cm of soil at 81 sites. Plasmodia, plasmodial tracks or fruiting bodies were used as evidence for the presence of myxomycetes in primary isolation plates, but only 26/81 sites yielded myxomycetes.

Stephenson and Landolt (1995) obtained data on the distribution and abundance of myxomycetes in soils of two study areas in eastern North America by recording the plasmodia appearing in the same hay infusion agar plates prepared for the isolation of dictyostelids. The impetus for this particular approach to studying soil-inhabiting myxomycetes related to the fact that myxomycete plasmodia had often appeared in the same isolation cultures as dictyostelids during the course of surveys of the latter group carried out in a number of study areas in eastern North America (Landolt and Stephenson, 1986) and elsewhere in the world. The isolation procedures used were essentially those described by Cavender and Raper (1965) for dictyostelids. Primary isolation plates were incubated under diffuse light at 20-25°C and each plate was carefully examined at least once a day for several days during the first week and at less frequent intervals thereafter. Since plasmodia typically appear somewhat later than dictyostelids in such plates, the latter have to be maintained for at least several weeks. Plasmodia appearing in these hay infusion agar plates were routinely transferred to culture plates prepared with CMA/2, inoculated with a suspension of *Escherichia coli* and incubated at room temperature (*ca.* 20-23°C).

Limitations of Existing Techniques

One of the critical steps in isolating dictyostelids and myxomycetes from soils involves breaking up the soil particles to suspend the amoeboid cells present. Mechanical agitation by means of a rotary shaker has been used in a number of studies (e.g. Kerr, 1994; Stephenson and Landolt, 1995) but this treatment results in a loss of recoverable dictyostelid amoebae, with losses increasing with increased shaking time (Kuserk *et al.*, 1977). Presumably, the same would be true for myxomycete amoebae. Apparently, this loss is due to the shearing action the particles exert on the amoebae, with the probability of an individual amoeba being 'hit' by a sufficient number of particles to cause its destruction increasing with time. As such, the use of milder methods (e.g. by gentle swirling or inversions of the sampling container) for mixing samples would seem highly desirable.

As has been pointed out, enumeration of myxomycetes as PFUs will invariably underestimate the number of cells present. Precisely what PFUs are is still uncertain, but circumstantial evidence points to their being wholly or predominantly amoeboflagellate cells (or possibly microcysts, their resistant form) rather than plasmodia (including sclerotia and macrocysts, the two resistant forms of this stage of the myxomycete life cycle). In apomictic species, one cell could in theory give rise to a plasmodium, but this is unlikely to occur for every cell present. If the species present are heterothallic, more than one viable cell of each species would have been present in the culture for a plasmodium to form. Theoretically, if cells of different mating types are present in equal proportions, on average three randomly chosen cells are required to give rise to a plasmodium. Consequently, any figures for myxomycete abundance generated in this manner would probably have to be multiplied by a factor of at least two or three (and possibly more) to reflect the actual numbers of myxomycetes present in a given study area. Moreover, because plate count methods are invariably selective (i.e. not all representatives of a particular group of organisms are recovered under a given set of cultural conditions), the figures obtained are probably better regarded as indices of abundance than as actual numbers.

A major problem associated with surveying the biotic diversity of myxomycetes in soils relates to the simple fact that many of the myxomycetes recovered as PFUs in primary isolation plates fail to produce fruiting bodies, and identification of myxomycetes is based almost exclusively on features of the fruiting body (Martin and Alexopoulos, 1969; Stephenson and Stempen, 1994). Since most species of myxomycetes do not easily complete their life cycle in culture (if indeed they can be cultured at all), this situation is not surprising. Nevertheless, it is sometimes possible to induce plasmodia appearing in primary isolation plates to produce fruiting bodies, using techniques similar to those described by Clark and Stephenson (1994) for *Didymium ovoideum*. If this is to be attempted, plasmodia should be transferred to culture plates prepared with 2% water agar or CMA/2 and these plates incubated at room temperature (*ca.* 20-23°C) and maintained in the laboratory by weekly subculturing and feeding with sterile oak flakes (Clark, 1980).

Suggestions to Overcome the Limitations of Techniques

Clearly, collecting and handling methods that minimize disturbance in soil structure and moisture content of the sample until it is suspended in liquid should be utilized whenever possible. Ideally, samples should be processed in the laboratory as soon as possible following collection because the propagules gradually die off. Rarer species of dictyostelids will be lost within a few weeks, and it seems likely that the same is true for myxomycetes. Keeping samples from being exposed to major fluctuations in temperature is a necessity. In general, it is best to keep soil samples as cool as is practical, but not lower than normal refrigeration temperatures of about 4°C. However, it should be noted that the amoebae of some tropical species of dictyostelids do not survive long at refrigeration temperatures. Whether the same is true for tropical species of myxomycetes is not known. Unless the technique of freezing soil samples to partition active and encysted forms (Kuserk *et al.*, 1977; Kuserk, 1980) is to be used, samples should not be subjected to freezing, since this kills any trophic cells present. Interestingly, some studies of myxomycetes that have applied this technique to soil samples have shown unexpected increases in PFUs after freezing. These increases have been attributed to the breaking of endogenous dormancy in a large population of encysted forms that are not recovered from fresh samples. Presumably, the shock of the freezing procedure is required to activate these encysted forms, a phenomenon similar to that found in certain other groups of soil organisms. This suggests that in any truly comprehensive, survey of the myxomycetes present in soils at a particular site, two analyses should be carried out, one performed on a fresh sample and the other on a frozen sample.

For both dictyostelids and myxomycetes, an effort to quantify numbers in relation to the unit volume of soil occupied rather than the weight of the sample warrants consideration, since the latter varies with mineral composition and moisture level. This factor becomes most important if there is a need to obtain comparable quantitative data from different sites and sampling is to be carried out by different workers.

Integrating Techniques into a General Scheme of Analysis

Any effective survey and/or inventory of the biotic diversity of the organisms present in soils at a particular site will involve the collection of a series of samples. As already noted, a small portion of each sample provides sufficient material for the isolation and enumeration techniques used for dictyostelids and myxomycetes. When this material is processed, aliquots obtained from the same subsample can be used for both dictyostelids and myxomycetes. Additional aliquots could provide the basis for qualitative and quantitative studies of other microorganisms (e.g. ciliates, myxobacteria, nematodes and filamentous fungi) present in the soil microhabitat.

Bibliography

General Texts

Alexopoulos, C.J. (1963) The Myxomycetes II. *Botanical Review* 29, 1-78. [A general overview of myxomycetes as organisms.]

Cavender, J.C. (1990) Dictyostelida. In: Margulis, L., Corliss, J.O., Melkonian, M. and Chapman, D.J. (eds) *Handbook of Protoctista.* Jones and Bartlett, Boston, pp. 88-101. [A summary of developmental studies, isolation and cultivation techniques, and taxonomy of dictyostelids.]

Hagiwara, H. (1989) *The Taxonomic Study of Japanese Dictyostelid Cellular Slime Molds.* National Science Museum, Tokyo. [A general reference which has descriptions of some species of dictyostelids not included in the Raper monograph.]

Martin, G.W. and Alexopoulos, C.J. (1969) *The Myxomycetes.* University of Iowa Press, Iowa City. [The standard monograph on myxomycete taxonomy.]

Martin, G.W., Alexopoulos, C.J. and Farr, M.L. (1983) *The Genera of Myxomycetes.* University of Iowa Press, Iowa City. [An abridged, but updated version of the Martin and Alexopoulos monograph, but with keys only to the genus level.]

Raper, K.B. (1984) *The Dictyostelids.* Princeton University Press, Princeton, 453 pp. [The single, most comprehensive monograph on dictyostelid taxonomy with sections on their history, ecology, cultivation and preservation.]

Stephenson, S.L. and Stempen, H. (1994) *Myxomycetes, a Handbook of Slime Molds*. Timber Press, Portland. [A relatively non-technical, but fairly comprehensive introduction to the biology, ecology and taxonomy of the myxomycetes.]

Specific References

Cavender, J.C. (1993) Response of soil dictyostelid slime molds to agricultural disturbance in a tropical environment. *Biotropica* 25, 245-248.

Cavender, J.C. and Raper, K.B. (1965). The Acrasieae in nature. I. Isolation. *American Journal of Botany* 52, 294-296.

Clark, J. (1980) Competition between plasmodial incompatibility phenotypes of the myxomycete *Didymium iridis*. I. Paired plasmodia. *Mycologia* 72, 312-321.

Clark, J. and Stephenson, S.L (1994) *Didymium ovoideum* culture and mating system. *Mycologia* 86, 392-396.

Feest, A. (1986) Numbers of myxogastrids and other protozoa recovered from Bohemian soils. *Ekologia* 5, 113-118.

Feest, A. (1987) The quantitative ecology of soil Mycetozoa. *Progress in Protistology* 2, 331-361.

Feest, A. and Madelin, M.F. (1985a) A method for the enumeration of myxomycetes in soils and its application to a wide range of soils. *FEMS Microbiology Ecology* 31, 103-109.

Feest, A. and Madelin, M.F. (1985b) Numerical abundance of myxomycetes (myxogastrids) in soils in the West of England. *FEMS Microbiology Ecology* 31, 353-360.

Feest, A. and Madelin, M.F. (1988a) Seasonal population changes of myxomycetes and associated organisms in four woodland soils. *FEMS Microbiology Ecology* 53, 133-140.

Feest, A., and Madelin, M.F. (1988b) Seasonal population changes of myxomycetes and associated organisms in five non-woodland soils, and correlations between their numbers and soil characteristics. *FEMS Microbiology Ecology* 53, 141-152.

Horn, E.G. (1971) Food competition among cellular slime molds (Acrasieae). *Ecology* 52, 475-484.

Indira, P.U. (1968) Some slime moulds from southern India. IX. *Journal of Indian Botany* 41, 792-799.

Kerr, S.J. (1994) Frequency of recovery of myxomycetes from soils of the northern United States. *Canadian Journal of Botany* 72, 771-778.

Kuserk, F.T. (1980) The relationship between cellular slime molds and bacteria in forest soil. *Ecology* 61, 1474-1485.

Kuserk, F.T., Eisenberg, R.N. and Olsen, A.M. (1977) An examination of the methods for isolating cellular slime molds (*Dictyosteliida*) from soil samples. *Journal of Protozoology* 24, 297-299.

Landolt, J.C. and Stephenson, S.L. (1986) Cellular slime molds in forest soils of southwestern Virginia. *Mycologia* 78, 500-502.

Stephenson, S.L. (1989) Distribution and ecology of Myxomycetes in temperate forests. II. Patterns of occurrence on bark surface of living trees, leaf litter and dung. *Mycologia* 81, 608-621.

Stephenson, S.L. and Landolt, J.C. (1996) The vertical distribution of dictyostelids and myxomycetes in the soil/litter microhabitat. *Nova Hedwigia* 62, 105-117.

Thom, C. and Raper, K.B. (1930) Myxamoebae in soil and decomposing crop residues. *Journal of the Washington Academy of Science* 20, 362-370.

Warcup, J. (1950) The soil plate method for the isolation of fungi from soil. *Nature* 166, 117-118.

Yeasts 9

Jack W. Fell[1] and Cletus P. Kurtzman[2]

[1]Rosenstiel School of Marine and Atmospheric Science, University of Miami, Key Biscayne, FL 33149, USA; [2]National Center for Agricultural Utilization Research, Agricultural Research Service, US Department of Agriculture, Peoria, IL 616104, USA.

The Role of Yeasts in Ecosystem Processes

Yeasts are a polyphyletic group of ascomycetous and basidiomycetous fungi whose unifying characteristic is a unicellular growth phase, typified by budding cells or by cell fission. There are approximately 101 known genera and 800 species of yeasts. Current estimates indicate that these species represent less than 10% of the species that occur in nature. Systematic criteria employ life histories, morphology and differential abilities to utilize a variety of carbon and nitrogen compounds.

Most yeasts are saprotrophs that occur in association with living and dead organic plant and animal material; some species are pathogenic, causing diseases in plants and animals (invertebrates and vertebrates). The occurrence of certain species of yeasts is dependent on specific physiological capabilities, such as the ability to utilize a certain compound or to survive and grow at low or high temperatures. For example, the occurrence of the yeast *Candida sonorensis* in cactus tissue is related to the yeast's ability to utilize propan-2-ol, which is a prevalent alcohol in cacti. Similarly, the psycrophilic yeast *Mrakia frigida* is found in Antarctic soils. Therefore, the occurrence of certain species of yeasts in soils is usually the combined result of environmental conditions and the presence of specific organic compounds. The latter is often a reflection of the associated plant and animal communities, although exogenous factors such as introduced organic matter (natural and pollutant) will affect the yeast community structure. Concentrations of yeasts in soils will vary from a few cells to tens of thousands per gram, although numbers as high as 250,000 cells per gram have been recorded in soils under fruit trees and berry bushes.

Yeasts are used in many industrial processes, such as the production of alcohol (beverage, industrial and fuel), biomass (baker's, food and fodder yeasts) and metabolic products (enzymes, vitamins, capsular polysaccharides, carotenoids, polyhydric alcohols, lipids, glycolipids, citric acid, ethanol and carbon dioxide). Yeasts are also a source of compounds synthesized by the introduction of recombinant DNA and they are used in the biological control of post-harvest diseases of fruits and grains.

The important commercial value of known species of yeasts leads to a considerable potential for the vast number of unknown species. The discovery of new species and strains raises a critical issue: yeasts and other fungi are of primary importance in recycling of organic material, particularly in subtropical and tropical regions, where there is a high diversity of species and a rapid rate of organic turnover. However, many of these environments are being degraded rapidly. As the organic constituents of these habitats change, so do the fungal populations, and consequently many species of yeasts are being lost. There is an urgent need to collect, isolate and characterize yeast communities.

Sampling, Isolation and Culture Methods

Sampling/Collection Methods

The majority of the yeasts, and other fungi, are obligate aerobes and require oxygen for growth and reproduction. They are usually found in the upper (aerobic) layers of soils and sediments. Exceptions occur, such as in ice cores, where yeasts can be preserved in deep layers of ice.

Soil and sediment samples can be manually collected from terrestrial and shallow water environments using sterile jars, vials, plastic bags ('Ziplocks') and coring devices. Deep water sediments can be remotely collected using grabs, gravity and piston cores or submersibles. The size of the sample can be a volume of *ca.* 50 ml and multiple samples can be taken from a single location or from the grab. Excess water does not need to be included with the sample, as the presence of water often promotes an anaerobic condition. Cores can be sectioned and samples taken from recorded depths along the core. Samples should be processed immediately or held on ice, if possible, until processing can be initiated. Maintaining low temperatures is particularly necessary when collecting from low-temperature habitats. Sample containers should not be airtight to avoid formation of anaerobic conditions, although the containers should be adequately covered to eliminate contamination.

Sample processing should be done in a contaminant-free environment, although this is not always feasible under field conditions. Where possible, a culture room or hood is preferable. It is imperative to avoid air draughts and inexpensive, temporary hoods can be fabricated for field use. Utensils may be sterilized with flamed alcohol and bench work space should be kept clean with soap or soap/bleach (e.g. 'Clorox') solution.

Isolation Methods

Direct sampling

Soil particles can be placed directly on an agar medium, allowed to incubate for 24 or more hours and then the particle is microscopically examined *in-situ* under low power for growth at the particle/medium interface. Suspected yeast colonies can be picked and transferred to a microscope slide for inspection. Confirmed yeast growth can be transferred from the isolation medium to a growth medium. Streak cultures should be made to ensure purity of the culture. Colonies resulting from single cells can be transferred to slant culture tubes for further study.

The soil dilution technique

A known quantity of soil can be placed in a test tube with a given volume of sterile water (seawater in the case of marine samples), vortexed and a 1:10 dilution made in fresh or seawater. Standard spread plates are prepared from each of the dilution series. Water may be filtered through 47 mm diameter nitrocellulose filters, 0.45 μm pore size, using a vacuum filter apparatus (such as that supplied by Millipore Corp.), and the filter placed face-up on solid nutrient agar. Culture plates are incubated, colonies enumerated and isolated after appropriate periods of time depending on incubation conditions.

Enrichment procedures

These consist of placing soil samples in flasks with a nutrient medium, followed by incubation with continuous oscillatory shaking for 12-24 h or more. The liquid is sampled by streak or spread plate. Shake culture usually forces the filamentous fungi to grow in balls, whereas the yeasts remain as free cells in the liquid. Such techniques are highly selective, as they may favour the growth of one species over another, and they cannot be used to enumerate natural populations.

Culture Methods

Isolation and incubation conditions

These should be designed to maintain ambient environmental conditions. For example, when examining samples from polar and other cold regions, all samples should be kept cold, on ice if necessary. The incubation temperature should be *ca.* 5°C. In contrast, sampling for *Cyniclomyces guttulatus,* which occurs in rabbit faeces, an incubation temperature of 35-40°C and an atmosphere enriched with 15% (v/v) CO_2 is required. Overgrowth by filamentous fungi is an omnipresent problem. In addition to the method described for shake cultures, samples may be incubated at temperatures lower than ambient; however, both these techniques are imperfect solutions to this problem. For temperate and tropical species, a temperature of 12°C usually slows down the growth rate of filamentous fungi. Culture plates must be examined daily to isolate the yeast strains as soon as they appear and before filamentous fungal overgrowth occurs.

Isolation of every colony which grows on an isolation plate will result in an overwhelming number of isolates. Therefore, colonies on agar plates should be classified on the basis of colour, size and morphology. Each type of colony should be counted and two to three representatives selected for isolation. Colonies are picked off with a finely tipped needle for transfer to a nutrient agar plate and streaked for purity testing. Single colonies from the purity plate are then transferred to a slant of yeast isolation medium (see later) in a culture tube. The strain can then be identified using standard yeast taxonomic methods (van der Walt and Yarrow, 1984; Yarrow, 1996).

Competition from bacteria

Lowering the pH of the isolation medium or adding antibiotics is usually employed. Most yeasts are tolerant of low pH and lowering the pH of the nutrient medium to 3.5-4.0 will inhibit the growth of some, but not all bacteria. Agar undergoes rapid hydrolysis when autoclaved at pH values greater than 4, and so, to avoid liquefaction, a predetermined volume of 1M HCl is added to sterilized molten agar after it has cooled to 45°C. Certain broad-spectrum antibiotics (e.g. chloramphenicol or a combination of penicillin/streptomycin), which inhibit bacterial growth or reproduction, but which do not affect yeasts, can be incorporated into the growth medium.

Design of the growth medium

The type of soil substrate and the type of yeast to be isolated should be considered. As already mentioned, *Cyniclomyces guttulatus* requires the presence of amino acids in the medium, whereas *Rhodosporidium lusitaniae* was isolated from woodland soil using media containing lignin-related phenolic compounds as the sole carbon source. Therefore, it is possible to select for specific types of yeasts by altering the growth medium using specific carbon and nitrogen compounds, vitamins and other growth factors which may be specific to the habitat, water activity using increasing concentrations up to 45-60% (for osmophilic species), salinity (for marine species), etc.

Many general media have been described. We favour yeast isolation medium (a liquid medium) containing (all w/v) 2.0% glucose, 1.0% peptone and 0.5% yeast extract. A widely used alternative is Wickerham's YM medium, which contains 0.3% yeast extract, 0.3% malt extract, 0.5% peptone and 1.0% glucose. Other carbon sources can be substituted for glucose (but must be autoclaved separately and added to molten, cooled agar to avoid thermal decomposition).

Freshwater or filtered seawater can be used with either medium. For oceanic sampling, seawater at 37°/oo is used, but in inshore and estuarine waters the salinity is reduced, usually to 15°/oo, or to suit local conditions.

Solid media are prepared by the addition of 2% agar. Antibiotics can be added to the media, e.g, chloramphenicol can be used at 200 mg l^{-1} (0.02%), which has the advantage that it can be added prior to autoclaving. An alternative is a mixture of penicillin G and streptomycin sulphate, added dry at the rate of 150-500 mg of each l^{-1} to autoclaved, cooled (45°C) medium.

Limitations of Existing Techniques

In designing collection methods, a basic ecological premise must be considered, specifically that distributions of organisms is usually patchy. Individual cells are not evenly distributed; rather they inhabit the soils and sediments in aggregates. In low-nutrient environments cells may be in a dormant or low metabolic state and then act opportunistically, growing and reproducing rapidly with the introduction of organic particulates or soluble compounds. Conversely, the organisms may be species that were introduced into the soil with (for example) the death of a plant or animal. These introduced species may or may not be able to survive in soils. Therefore, results from one sample taken at one time period cannot be extrapolated to other times and locations (even a few centimetres distant).

Another point of concern is that all isolation media are selective; in addition, certain rapid growing ubiquitous species may be the dominant strains to grow on artificial media. The strains that are prevalent in soils may not grow and reproduce outside their natural substrate. Plating techniques provide a further limitation due to the maximum number of cells (*ca.* 300) that can be accommodated on a single plate. If a rare species represents 1% or less of the community, it is possible that these species will not be selected, or they will be overgrown by weed species or overlooked. Additionally, cell numbers represent the standing crop of a given population and do not reflect factors such as turn-over rates, hyphal fragmentation, spore release or rates of consumption by various invertebrates. As a result, community structure analysis may not represent the natural populations.

Suggestions to Overcome Limitations of Techniques

As discussed above, careful consideration of the environmental and soil/sediment conditions is necessary. A variety of media and incubation conditions should be employed, followed by close and repeated inspection of the culture plates to ensure that representatives of all species have been identified.

Multiple collections at each site are required to delineate a representative distribution pattern and community structure.

Integrating Techniques into a General Scheme of Analysis

Collections must be designed in conjunction with other disciplines in order to achieve an optimal understanding of the interactions with other organisms. Sufficient environmental data should be collected to understand the site ecology. Through the use of the Global Positioning System (GPS) it is possible to accurately locate the site, and the environmental conditions/ecology should be described in detail, due to current (and rapid) environmental degradation. Records should include type of habitat, plants and animals, soil type (sandy, loam, etc.) and organic content, elevation, weather, date and time of day. Samples taken from bodies of water can include water depth, temperature, salinity (if appropriate), pH and oxygen concentration.

Bibliography

Parkinson, D., Gray, T.R.G. and Williams, S.T. (1971) *Methods for Studying the Ecology of Soil Micro-organisms.* IBP Handbook No. 19, Blackwell Scientific Publications, Oxford. [Chapters on bacteria and fungi, habitat description, sampling, determination of the form and arrangement of microorganisms in soil, isolation of microorganisms, biomass measurements, determination of microbial activity in soil, identification of soil organisms, media for isolation of soil microorganisms.]

Phaff, H.J., Miller, M.W. and Mrak, E.M. (1966) *The Life of Yeasts. Their Nature, Activity, Ecology and Relation to Mankind.* Harvard University Press, Harvard. [A small and very readable book that presents the basic biology of yeasts.]

van der Walt, J. P. and Yarrow, D. (1984) Methods for isolation, maintenance, classification and identification of yeasts. In: Kreger-van Rij, N.J.W. (ed.) *The Yeasts. A Taxonomic Study*, 3rd edn. Elsevier, Amsterdam, pp. 45-103. [Detailed methods and media for the above topics. The standard reference for yeasts.]

Yarrow, D. (1996) Isolation, maintenance and identification of yeasts. In: Kurtzman, C.P. and Fell, J.W. (eds) *The Yeasts. A Taxonomic Study*, 4th edn. Elsevier, Amsterdam (in press). [An updated version of van der Walt & Yarrow, 1984.]

Zoosporic Fungi 10

Geoffrey S. Hall

International Mycological Institute, Bakeham Lane, Egham, Surrey TW20 9TY, UK.

Zoosporic Fungi in Terrestrial and Aquatic Ecosystems

Zoosporic fungi are a polyphyletic assemblage which includes organisms currently classified with the *Fungi* (the *Chytridiomycetes*), with the *Chromista* or *Straminipila* (the *Oomycetes*, *Hyphochytriomycetes* and *Labyrinthulomycetes*) and with the *Protozoa* (the *Plasmodiophoromycetes*). This group also has several genera of uncertain taxonomic status which are included with the *Protoctista*. General accounts on the ecology and systematics of this group are given by Sparrow (1960), Fuller *et al.* (1964), Jones (1976) and Fuller and Jaworski (1987).

Zoosporic fungi are a highly diverse group and include species adapted to aquatic, amphibious and terrestrial environments. Disturbed land and moist habitats are favoured habitats and the majority of known species are found in moist soil and freshwater habitats, especially around lake margins. Although they are most abundant in more or less neutral soils, they have also been recovered from acid peat and highly alkaline soils and have a worldwide distribution. Terrestrial species are most common in pasture and other grasslands, but are less common in dry woodland, although some species have even been found in sand in canyons in the Arizona and Mexican deserts. Relatively few species have been described from marine and brackish water environments, and most of these are from decomposing algal and plant remains (especially in coastal mangrove forests in the tropics) with a few others in estuarine and near-shore sediments. However, they may be isolated from nearly any piece of organic debris or biological surface in estuarine or coastal waters, and so the relatively small number of described species is more likely to be a reflection of low sampling intensity. They appear to be more common in estuarine environments than off-shore and in the open ocean, and have not yet been recovered from benthic sediments.

Most are saprotrophs, living on decaying plant and animal debris, including fruits, twigs, insect and crustacean cadavers, arthropod exoskeletons and keratinous substrates. Others are facultatively or obligately biotrophic parasites of a wide range of animals, plants and fungi (mycoparasites). Hosts include insects, freshwater and marine crustaceans, rotifers, nematodes, fish and mammals, many freshwater and marine algae (uni- and multicellular species), terrestrial, salt marsh, marine and freshwater green plants and fungi (especially the hyphae and oospores of other zoosporic fungi and mycorrhizas). Some species have complex life cycles and alternate between dipteran larvae and copepod and ostracod alternate hosts. Although most species depend on water (including rain splash) for propagule dispersal, a few (not treated here) use air for dispersal.

Their major role in terrestrial, freshwater and brackish ecosystems is in the degradation and recycling of nutrients. In marine environments, nearly all the described species are parasites, especially of algae and crustaceans, although a few species feed heterotrophically, grazing on bacteria and yeasts. Therefore, they have an important role to play in regulating populations of these organisms.

Although most are obligate aerobes, some species are facultative anaerobes and can tolerate the extremely low oxygen and elevated carbon dioxide levels which often occur in stagnant water. Recent research has shown that some species of these fungi, previously thought to be protozoa, are obligate anaerobes and break down cellulose in the guts of ruminant mammals. Many species can utilize a wide range of carbohydrate and nitrogen sources, although a small number (mostly mycoparasites) require organic nitrogen sources. Some species are auxotrophic for vitamins and sterols, whereas others are obligate biotrophs. Many are highly intolerant of pollution and are particularly sensitive to heavy metals, especially copper, although some marine species are most often encountered in algae growing in heavily polluted water. Some species are obligate halophiles, but others can tolerate a wide range of salinity, a change in the concentration of which may be required for sporulation to occur.

Some are important pathogens of a wide range of economically important crops (mono- and dicotyledons) and occur in temperate, subtropical and tropical climates. They principally cause root, stem and collar rots, bud and fruit rots, producing heavy crop losses in some years. Fish, especially Salmonids, and a wide range of crustacean shellfish including their larvae and eggs are also infected. Only one species, *Pythium destruens*, is pathogenic to mammals, infecting horses, dogs and possibly human.

While some are cosmopolitan, occurring on many hosts, others have restricted geographical and/or host ranges. Plurivorous species can have devasting impacts on natural environments. For example, *Phytophthora cinnamomi* has already destroyed over 10,000 hectares of native forest in Western Australia.

Sampling/Extraction Techniques

Soil Collection and Treatment

Soil samples from ditches, ponds, river banks, arable land, forests, deserts, sand dunes, dried river beds and composts can all be used. However, not all contain a highly diverse assemblage of organisms. The most diverse communities occur in rural ditches, dry depressions and vernal ponds which are intermittently dry, but moist soils from wooded areas, the edges of ponds, streams, lakes and rivers or sediment and debris from freshwater sites may all yield many species. Soils beneath moss mats, around roots of vegetation and in bogs have relatively fewer species, but some of these are very uncommon and are restricted to these habitats. Sandy soils have relatively less diverse communities and soils polluted by heavy metals have almost none.

At least 5 g soil should be collected and examined several times: 100 g should be collected and examined many times for rarely found species. Soils may also be sampled using a 1 cm diam. corer, inserted to a depth of *ca.* 5 cm (Dick and Newby, 1961). Fresh soils should be used within 6 hours, but can be stored for 3-4 weeks, if kept cool. Freshwater aquatic sediments and wet soils should be dried to the extent that they do not become anaerobic in storage before isolation. Insect exuviae are often present in sediments, and are best collected by placing a large quantity of wet debris from streams of lakes, either found along the shore or attached to reeds, in shallow enamel dishes. The debris is mixed with a small amount of water and the dishes gently rotated so that the individual exuviae become detached from the clumps of debris and float free on the surface.

Baiting

This is the most widely used and reliable technique to isolate zoosporic fungi from samples of soil or sediment and from associated substrates. The purpose of baiting is to substantially increase the population of fungi so that they can be studied in a condition resembling their natural state and to provide suitable stages of development for isolation and identification. Baiting is used because zoosporic fungi in soils produce sporangia when flooded with water which release motile zoospores which swim to the bait and colonize it - the soil-water-bait system. After colonization, the bait is then removed and plated out onto selective media to isolate the fungi growing in it.

Soil-water plates

Usually about 0.5 g sample soil or sediment is placed in a 9 cm diam × 2 cm deep Petri dish and the dish is filled to two-thirds its depth. It is important to ensure that the supply of oxygen to incubation vessels is not restricted and so deeper vessels should not be used. The contents are thoroughly mixed and the particulate fraction is allowed to settle out. Culture fluids should reflect the source of the sample, and are usually sterile distilled water, but autoclaved pond or river water, autoclaved seawater or 4-10% (w/v) NaCl solution are also used to suit the origin of the sample. Soils and sediments with very high organic matter contents (e.g. rainforest soils and composts) can be diluted by mixing with sterile sand (1:15) or adding 5 g to 1l distilled water. However, if desert or sand dune soils with a low organic matter content are being examined, up to 2 g can be added to one plate. Soils often yield different fungi following drying, as competing organisms (e.g. protozoa and some higher fungi) are killed or suppressed. It is a good idea to split a sample before study and to air-dry half at room temperature for 2-3 months, or to decant the fluid from already baited soil-water plates and air-dry the residue as before, and then flood and bait this soil again.

Types of bait

The type of bait chosen determines the type of fungi isolated, i.e. baits are selective. They consist of a wide variety of organic substrates and the most useful baits should be susceptible to colonization, have a high selectivity and be readily available. Seeds are most commonly used, but a variety of fruits, leaves and seedling roots can also be used.

Seeds of hemp (split and boiled in several changes of water to remove as much of the oil present as possible), grass, maize or sesame (hulled) are either placed on a gauze overlying saturated soil, or directly into a soil-water suspension. Usually two are placed on the bottom of the plate and two are floated on the surface. Good for most saprotrophs.

Pollen (especially pine pollen) may be used fresh or preserved, although this influences the range of species isolated. It can be stored dry indefinitely or can be dried at 60°C or irradiated to preserve it. Good for unicellular species.

Algae can be used live or killed by heating for a few minutes at 70°C (do not boil). Various freshwater algae, e.g. species of *Spirogyra, Cladophora, Vaucheria, Tribonema,* or marine algae, e.g. diatoms, small sterilized pieces of green, red or brown algae may be used. Good for many unicellular species and some parasites.

Cellulose (as 'Cellophane' or dialysis tubing) cut into small strips, washed to remove plasticizers and dried is used to isolate cellulolytic species. As it is translucent, mounts can also be made to examine the organisms isolated *in situ* under the microscope. Sterile filter paper is often used and is good source of cellulose, but is opaque.

Translucent insect wings (fly, cockroach, etc.) or decalcified prawn or shrimp exoskeleton are used for chitinolytic species. Autoclaved brine shrimps are required to isolate a few marine species.

Snake skin (obtainable from your local zoo, pet shop or herpetologists' club), cut into 1 cm squares and boiled, then floated on the surface, or defatted human hair may be used for keratinolytic species.

Fruits, such as small apples, pears and tomatoes, are used for plant pathogens. Fruits should be rinsed vigorously under running tap water to remove bacterial masses and then examined in water with a stereomicroscope. Rose hips are a good bait for some saprotrophs. In addition, some plant pathogens may be isolated by boring 1 cm diam holes in fruits (apple, pear, mango, papaya, etc.), filling them with 0.5 g soil and incubating them in humid conditions (e.g. in a plastic bag) in an incubator until a soft rot develops. The tissue is excised and plated onto selective media.

Surface-sterilized seedlings or the excised radicles of lupins, other legumes, grasses or other suitable plants are floated on water over a layer of soil. Brown or orange lesions develop behind the root tip which contain pathogenic species. Good for plant pathogens and some saprotrophs.

Small portions of fresh leaves, 1-2 cm wide, cut from leaves of *Citrus, Hebe, Chamaecyparis* or tomato, or basal segments or whole leaves of pine needles floated in water above soil samples are useful for many plant pathogens and some saprotrophs.

Autoclaved grass blades, either fresh or dead from old plants (to render visible structures which form inside the leaf blade), are placed in dishes of water with some infected material opposite. Good for saprotrophs and some plant pathogens.

A balance between the amount of material added and the volume of water should be maintained to avoid fouling by excessive bacterial growth. Baits should be examined every 2-3 days over a 21-28 day period and fresh or different baits can be added when the original ones become colonized. Some slow-growing species or those which develop from germinating oospores can take up to 10 weeks to appear and so plates should be retained and baited over this period if it is desirable to obtain the maximum diversity in a soil or sediment. Baits are then removed and plated onto selective media.

Isolation of Fungi from Baits

Baits are removed, gently washed with distilled or de-ionized water and wet mounts are examined under a stereomicroscope. Individual species are recognized by distinctive morphological features. Under the stereomicroscope, several hyphae are removed using fine-pointed forceps or a mounted needle, separated and several single hypha are placed around the periphery of a suitable agar medium (Fig. 10.1).

The remainder of the bait can then be transferred to another dish of sterile distilled water or plated out onto agar. Isolation plates should be examined after 24 hours and small colonies excised with a scalpel tip and transferred to fresh media. Intracellular fungal parasites may also be isolated with these hyphal cultures and manifest themselves by aberrant hyphal growth, usually swellings in the hyphae caused by the production of sporangia or sexual spores.

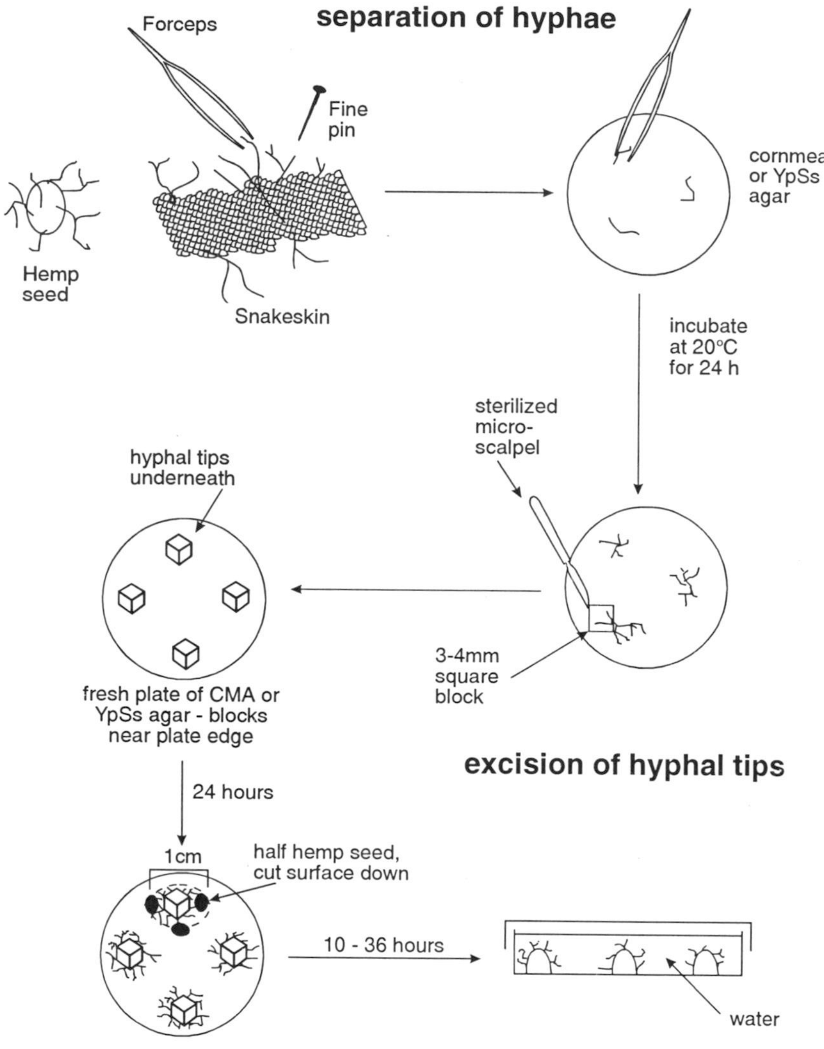

Fig. 10.1 Method for the isolation of zoosporic fungi from baits.

Sporangia of many chytrids are detached from a small amount of infected material in a plate of suitable agar medium and moved across the plate to separate them from the source material, which is then removed by excision or killed with a drop of alcohol. A high-magnification dissecting microscope with a good working distance and fibre optic lights ('cold illumination') are essential. Transfer to fresh agar should be done within 2-3 days to avoid overgrowth by contaminants, when growth can be encouraged by adding a drop of fresh water. Some cultures grow better in broth culture and this should be tried for cultures which become weak or which fail to grow on agar. Resting spores of chytrids may take 10 days or more to form and so the cultures should be retained. Some more complex methods for the isolation of chytrid sporangia which require patience and great manual dexterity are given in Couch (1939).

Isolation of Fungi from Debris

Soil and sediment samples also contain woody substrates, such as waterlogged twigs, which harbour several genera of chytrids and *Oomycetes*. Small sections, 5-15 mm in diameter, of corticated twigs from aquatic sediments or soils can be flooded with distilled water in a Petri plate and incubated for 10 days until fungi are seen. The twig culture should be continued for 30 days and the base examined for other species which typically grow on the base of the Petri plate. Alternatively, twigs are packed into jars, filled with distilled water, one or two baits (cherry tomatoes or rose hips) are added and the jar is sealed. When pustules of fungi appear on the baits, they are removed and examined. This method selects for facultatively anaerobic species. Waterlogged leaf pieces from aquatic sediments can be removed, washed to remove bacteria, placed on a Petri plate and baited with hemp/sesame seeds. Any root tissues present in the material should be washed thoroughly, placed in pond water or saline, as appropriate, incubated and examined for sporangia. Also, pieces of marine plants or debris can be dragged across the surface of an agar plate, which often gives discrete colonies. When zoospores are produced, the water may be baited as previously described.

Selective Isolation Media

These are sometimes employed to isolate fungi directly from soils or sediments. Chitinolytic or cellulolytic organisms may be isolated directly by dilution plating onto specific media which contain N-acetyl glucosamine, cellulose or xylan agar, but it is often easier to encourage fungi to grow on baits in gross culture and build up their population and then transfer ripe sporangia to an appropriate medium. Various antibiotic chemicals are added to the isolation medium to inhibit the growth of bacteria and fungi (e.g. 0.02 g l^{-1} 'Benomyl' and 0.6 g l^{-1} each of Penicillin G and Chloramphenicol) or VP_3 medium (see Appendix). Fast-growing species which may overgrow many slower species may be limited by the addition of hymexazol (3-hydroxy, 5-methyl isoxazole). Sodium chloride or sea salt, 10% (w/v), should be included when isolating and growing estuarine species. Media become acid with growth, which may prevent alkalophilic organisms from growing, but most buffers are toxic to many chytrids and even NaOH and HCl reduce growth, and this problem is not easily overcome.

Dilution Plating

This technique is sometimes used to isolate from soils, but is most effective when used with a selective medium (e.g. cellulose agar). A dilution series in either 2% (w/v) peptone water or 10% (w/v) saline is made to $1{:}10^{-4}$ and aliquots of either 0.1 ml are spread onto solid media (spread plates), or of 1.0 ml are added to cooled liquid medium held at 45°C, dispersed, poured and left to cool (pour plates).

Incubation Conditions

The temperature, pH and osmotic potential must reflect the environment from which the soil or sediment was collected. Zoosporangium formation and zoospore release are particularly sensitive to environmental conditions. Although samples are often left at room temperature, fluctuations are too great and it is better to place the cultures in an incubator, preferably with a light cycle. Incubation should be done at an appropriate temperature which reflects the origin of the species, e.g. 10-15°C for cool temperate species, 20-25°C for predominantly tropical species and 30-45°C for material from unusual high-temperature springs and other extreme environments. It is often a good idea to incubate soil-water plates at 15°C and at 30°C to be sure of isolating as many species as possible. There appears to be no real difference between the numbers of propagules recovered between the two techniques, although pour plates allow a greater volume of the sample to be examined (Hallett and Dick, 1981).

Removal of Contaminants

Often primary isolations are heavily contaminated with bacteria which grow in the surface mucilage surrounding hyphae; contaminating fungi are seldom a problem. Growing tips of hyphae are excised and plated onto isolation media with added antibiotics, e.g. 500 mg l^{-1} of Penicillin G with 500 mg l^{-1} Chloramphenicol as a general isolation medium. Streptomycin 200-500 mg l^{-1} and Neomycin 200 mg l^{-1} and/or Chlorotetracycline 10-20 mg l^{-1} may be used instead, but each inhibits the growth of some species.

Alternatively a modified 'Raper ring' may be used. These are made by folding a short strip of aluminium foil, about 5 mm wide, into a ring which is heated in a flame for 10 s until sterilized, and then placed onto a Petri plate of agar. The hot ring melts the agar and it sinks, ideally to about 2 mm deep. When the ring has cooled and the agar solidified, the plate is inoculated in the centre of the ring. Hyphae grow down under the ring and up outside it, leaving bacteria trapped in the middle on the agar surface. Several repetitions may be necessary until some cultures are completely free of contamination. To check for purity, cultures should be incubated at an appropriate temperature in nutrient broth, when any bacteria present will cause turbidity.

Culture Media (See Appendix)

Generally, dilute media are much better than those with high levels of nutrients and encourage sporulation rather than excessive mycelium production. Weak V8-juice agar (1:1), corn meal agar or bean/pea deconcoction agar are good general media for laboratory studies. Corn meal agar is also translucent, allowing observation with the microscope on the Petri plate, and V8 agar has a soft consistency, enabling slides to be made easily. Dilute (1:1) corn meal agar and other weak media are useful for isolates of many genera of *Oomycetes*, whereas GYPS agar is best for *Saprolegnia* and many chytrids. Some isolates produce sex organs in agar cultures, others require sterols, and sometimes 30 mg l^{-1} β-sitosterol or cholesterol is added to stimulate their production. Many chytrids and some Oomycetes (*Phytophthora* and mycoparasitic species of *Pythium*) require thiamine in the medium, but only its components, and so it can be added before sterilization (which destroys the molecule). Although many species can utilize inorganic nitrogen as ammonium or nitrate, some species (*Phytophthora* and mycoparasitic species of *Pythium*) require an organic source, such as tryptone, peptone or L-asparagine. Although many chitinophilic and keratinophilic species can be isolated from appropriate baits, many of these species can grow on non-specific agars in the laboratory. Media should be made up with distilled water for soil and freshwater fungi, but with seawater for marine species.

Some media are useful for specific groups of fungi, e.g. YpSs (Emerson's yeast-phosphate-soluble starch medium), Xy and NADG are used for chytrids; dilute glucose-gelatine hydrolysate seawater-based medium (KMV) for some marine thraustochytrids; and low-nutrient seawater-based media, such as serum seawater agar (SSA) for labyrinthuleans. Facultative anaerobes need special culture conditions involving continuous neutralization of the acid produced (Emerson and Cantino, 1948).

Examination of Isolates

All unifungal, bacteria-free cultures must be incubated in an aqueous environment to produce zoospores, the physico-chemical characteristics of which should reflect the original habitat of the organism being cultured. Often the same baits that were used to isolate the fungi are essential to induce asexual and sexual structures to form in culture.

For many freshwater species small agar blocks of cultures are placed in dishes of sterilized filtered pond water containing half a sterilized hemp seed. Mycelium radiates out from the seed and can be cut into squares by rocking a scalpel with a curved blade through it, placing a cover slip underneath the cut mycelium and carefully lifting it up. The cover slip is then inverted and the mycelium mounted.

For others, especially species of *Pythium*, a 5 mm square block is cut from the agar plate, placed in a Petri plate, a few autoclaved grass blades arranged over it, and covered with 1-3 ml of pond/river water. The plate is incubated at 15-20°C for 24 hours, or until hyphae have grown onto the grass blades, and then it is flooded with 10 ml sterilized pond water. Alternatively, the grass blades may be placed on the agar surface for 24-48 hours and then removed and placed in sterilized pond water. Incubation is continued until sporangia are formed, which may take 1-3 days, and so repeated observation is required. Chilling the cultures for 30 min at 4°C may help to induce zoospore release in some species.

For species of *Phytophthora*, no grass is necessary and a small piece of culture is placed in sterilized pond or river water (or a weak salt solution) and incubated until sporangia form. A few drops of non-sterile soil extract may help to induce formation of sporangia in some species.

Some estuarine organisms, e.g. species of *Halophytophthora,* need to be grown in salt water followed by a general dilution of salinity from 20‰ to 10‰ or to 0‰ (freshwater) before sporangia will form. Others need to be grown in seawater, dehydrated for 2 hours and then rehydrated in fresh water. Often simultaneously decreasing the temperature by 5°C and increasing the amount of light available enhance this process.

Oogonia of homothallic species form in single cultures and may be seen through the reverse of the agar plate or on the external mycelium, in the agar block or in the grass blades in slide preparations. In heterothallic species of *Pythium, Phytophthora* and *Achlya*, oospores only form when mated with an opposite compatibility type, which may be of the same species (intraspecific pairing) or another species (interspecific pairing). Successful mating is sometimes difficult to achieve, especially with cultures of *Pythium* spp.

Specialized Techniques

Algal parasites

Endobiotic, holocarpic parasites of diatoms and other algae are difficult to isolate and maintain in culture, although some monoxenic (dual-membered) cultures have been established (Chakravarty, 1970; Raghu Kumar, 1978).

Marine Labryinthulomycetes

This group must be maintained in dual culture with marine isolates of the yeast genera *Rhodotorula* and *Torulopsis*, on which they feed.

Mycoparasites

May be isolated by placing 0.5 g fresh or air-dried soil on the youngest part of a plate of potato dextrose agar pre-colonized with a susceptible fungus (e.g. *Phialophora*) and scanned for the presence of oogonia after 7 days at 25°C (Foley and Deacon, 1985).

Direct Observation

Freshly collected filamentous or planktonic algae and all organic material (living and dead) should be examined, as interesting species may be found on rotifers, nematodes, insect exoskeletons and plant roots. Direct observation is also necessary for those species which are sensitive to changes in environmental conditions. Most of the known marine zoosporic fungi are found as parasites on algae and crustaceans. Cultures of fungi obtained for internal parasites and external parasites on oospores usually yield more kinds of zoosporic fungi.

Limitations of Existing Techniques

Some isolates fail to form sporangia and/or sexual structures in culture and can either not be identified or only be referred to a genus. Many isolates rapidly lose their ability to form zoospores and sexual structures, becoming wholly mycelial in culture. Some species of chytrids are obligately anaerobic and their isolation requires the use of sophisticated techniques in a nitrogen chamber. All these techniques are good for qualitative or distributional studies, but fail to give quantitative estimates of numbers and calculations of density or abundance are not possible. Isolation frequencies have sometimes been referred to as 'relative availability' of species, i.e. the relative numbers of species, or propagules of a single species, that can be induced to colonize the substrate used. For a fuller discussion of the problems of quantifying zoosporic fungi see Hallett and Dick (1981). A method for the examination of zoospores, based on the most probable number (MPN) technique was devised by Cooper (1993).

Suggestions to Overcome the Limitations of Techniques

Following isolation, rapid examination is essential to ensure that degeneration to a wholly mycelial state occurs. Molecular methods, especially the development of specific gene probes, have great promise for the detection of species in samples from the environment. Continuous-flow centrifugation, coupled with pour plating and scanning, can be used to estimate zoospore density (Hallett and Dick, 1981).

Integrating Techniques into a General Scheme of Analysis

Fractionation of organic debris by mild centrifugation, animals and plants may help in the isolation of species. As the zoosporic fungi occur in association with many other organisms, especially as parasites, these need to be examined as well.

Bibliography

General Texts

Fuller, M.S. and Jaworski, A. (1987) *Zoosporic Fungi in Teaching and Research.* Southeastern Publishing Corporation, Athens, Georgia. [An excellent summary of methods for many groups of zoosporic fungi.]

Fuller, M.S., Fowles, B.E. and McLaughlin, D.J. (1964) Isolation and pure culture of marine phycomycetes. *Mycologia* 56, 745-756. [Good, general text].

Jones, E.B.G. (1976) *Recent Advances in Aquatic Mycology.* Elek Science, London. [Especially the chapters The Ecology of Marine Lower Fungi by G.B. Bremer and The Ecology of Aquatic Phycomycetes by M.W. Dick.]

Sparrow, F.K.Jr. (1960) *Aquatic Phycomycetes,* 2nd edn. University of Michigan Press, Ann Arbor. [Isolation methods: pp. 21-38. The standard reference work on this group, although its systematic content is now dated.]

Specific Texts

Chakravarty, D.K. (1970) Production of pure culture of *Lagenisma coscinodisci* Drebes parasitizing the marine diatom *Coscinodiscus. Veröffentlichungen des Institutes für Meeresforschung Bremerhaven* 12, 305-312.

Cooper, J.A. (1993) Estimation of zoospore density by dilution assay. *The Mycologist* 7, 113-115.

Couch, J.N. (1939) Technic for collection, isolation and culture of chytrids. *Journal of the Elisha Mitchell Science Society* 55, 208-214. [A useful, but difficult to perform technique.]

Dick M.W. and Newby, H.V. (1961) The occurrence and distribution of *Saprolegniaceae* in certain soils of south-east England. I. Occurrence.

Journal of Ecology 49, 403-419. [Study includes isolation and sampling protocol.]

Emerson, R. and Cantino, E.C. (1948) The isolation, growth and metabolism of *Blastocladia* in pure culture. *American Journal of Botany* 35, 157-171.

Fell, J.W. and Master, I.M. (1975) Phycomycetes (*Phytophthora* spp. nov. and *Pythium* sp.nov.) associated with degrading mangrove (*Rhizophora mangle*) leaves. *Canadian Journal of Botany* 53, 2908-2922.

Hallett, I.C. and Dick, M.W. (1981) Seasonal and diurnal fuctations of oomycete propagule numbers in the free water of a freshwater lake. *Journal of Ecology* 69, 671-692.

Raghu Kumar, C. (1978) Physiology of infection in the marine diatom *Licmophora* by the fungus *Ectrogella perforans*. *Veröffentlichen des Institutes für Meeresforschung, Bremerhaven* 17, 1-14.

Appendix

Culture Media Formulations (See also Fuller and Jaworski, 1987)

V8-JUICE AGAR

V8-juice (Campbell's Soup Co.)	200 ml
Distilled water	800 ml
$CaCO_3$	2 g
Agar	15g

Dilute the juice with the water, then add the $CaCO_3$ and mix on a stir plate. Add the agar and autoclave. This will give an opaque medium. If a clear medium is required, add 5 g $CaCO_3$ to 354 ml of V8-juice, mix and centrifuge at 4000 rpm for 20 min. Decant the solution, dilute 1:4 (v/v) with distilled water, add 15 g agar l^{-1} and autoclave. (This is also called V8C medium).

YPSS AGAR

Soluble starch	20 g
Yeast extract	8 g
K_2HPO_4	1 g
$MgSO_4$: $7H_2O$	0.5 g

Make up to 1 l and autoclave. The starch and yeast extract should be put into solution by heating before adding the other ingredients, stirring continuously.

GYPS AGAR

Glucose	5.0g
Mycological peptone	0.5g
Yeast extract	0.05g
KH_2PO_4	0.5g
$MgSO_4$:$7H_2O$	0.15g
Agar	15.0g

Distilled water to 1 litre.

VP_3 AGAR

Corn meal agar	17 g
Agar (Oxoid no 1)	23 g
Sucrose	20 g
$MgSO_4$:$7H_2O$	10 mg
$CaCl_2$	10 mg
$ZnCl_2$	1 mg
$FeSO_4$:$7H_2O$	0.02 mg
$CuSO_4$:$5H_2O$	0.02 mg
MoO_3	0.02 mg

Make up to 990 cm^3 with distilled water and autoclave. When cooled to 50°C, add 100 mg PCNB (see below) aseptically and mix thoroughly to disperse.

PCNB

Vancomycin	75 mg
Penicillin G	50 mg
Rose bengal	2.5 mg

It is easiest to keep 10-fold stock solutions of the inhibitory agents, diluting these when adding to this medium. Plates must be stored in the dark and used within 36 hours.

KMV AGAR

Glucose	100 g
Gelatin hydrolysate	100 mg
Yeast extract	10 mg
Peptone	10 mg
Agar	1.3 g
Seawater	100 ml

Make up to 1 l and autoclave.

For all these media, autoclaving at 121°C for 15 min is a suitable method of sterilization.

Filamentous Fungi 11

Paul F. Cannon

International Mycological Institute, Bakeham Lane, Egham, Surrey TW20 9TY, UK.

Filamentous Fungi in Ecosystem Processes

Fungi play many roles in ecosystem function, but the most significant of these is decomposition of organic matter, unlocking nutrients and trace elements in dead plant and animal material for use by succeeding generations. They frequently represent the most significant group of soil organisms in terms of biomass and productivity (Kjøller and Struwe, 1982). Fungi are involved in a wide variety of mycorrhizal associations with plant roots (see Chapter 12), and many fungal pathogens are transmitted through soil. They are food for a wide variety of animals, and many are themselves parasitized or degraded by other fungi. Their economic significance as pathogens, mycorrhizae and metabolite producers is enormous.

Fungi can be divided in ecological terms into those which complete their life cycles within the soil, and those for which life in the soil is a part of a more complex system involving infection of aerial parts of plants, or of macrofauna. In these cases, fungi may either exist in soil as dormant propagules, or live saprobically on decaying host material.

Soil fungi are diverse for many reasons, which include the enormous variety of the fungal kingdom (Hawksworth, 1991), the specificity of many fungi to narrowly defined ecological niches, and the variation of soil microhabitats in spatial and temporal terms (Wainwright, 1988). Studies by Christensen (1989) of soil communities in Wyoming showed no clear evidence of completeness in sampling fungal species, even after analysis of well over 1000 isolates; as many as 286 species were identified. There are many other studies in which considerable diversity of species in soil samples (frequently well over 200) are charted (e.g. Gams and Domsch, 1969; States, 1978; Rambelli *et al.*, 1983). The large numbers of species present and the variety of their nutritional requirements mean that sampling strategies are often complex and must occur over a considerable period of time.

Sampling/Extraction Techniques

A large number of fungi produce fruit bodies on the soil surface, allowing detection and identification relatively easily. Most of these species are large saprobic or mycorrhizal basidiomycetes, but a number of groups of ascomycetes are also typical of the soil surface mycota. Removal of the fruit bodies (ensuring that the basal parts are removed intact) and air-drying is sufficient in most cases to provide material for identification. It is important to record field observations especially for fleshy taxa, particularly details of colour, latex production etc.

Direct observation of fungi below the soil surface is very time-consuming (Frankland *et al.*, 1990) and is rarely valuable for the purposes of diversity estimation, as most species are present either as vegetative mycelium or individual spores. Few species can be identified from this fragmental evidence, although samples from the litter layer and occasionally from large buried pieces of wood or other plant material may show more complete structures. These can also be incubated in damp chambers using moist filter paper or 'Vermiculite' (an expanded form of silica) as a water source (e.g. Keyworth, 1951). Mycelial cords can also be treated in this manner. For a large proportion of fungal taxa, isolation into pure culture is required, and techniques usually involve the direct processing of soil samples. Warcup (1955) detailed methods of culturing fungi from vegetative hyphae removed individually from samples, but such techniques are immensely laborious and have proved difficult for others to repeat (Frankland *et al.*, 1990).

Sampling techniques for diversity studies can be divided into general isolation protocols which aim to survey a wide range of soil fungi, and selective methods for study of specific taxonomic or ecological groups, which might otherwise be under-represented due to incorrect growth conditions or competition from wide-spectrum saprobes.

Isolation into pure culture may lead to a considerable diversity of fungal species sampled, but only a proportion of these will sporulate, and thus be identifiable using current technology. They may, however, be included in diversity studies without formal identification, as gross morphological features may be sufficient to allow separation of cultures into unnamed taxa (Hall, 1987). Bills and Polishook (1994) found that around 25% of fungal taxa isolated from litter fragments did not sporulate using standard protocols.

Many techniques have been developed to encourage maintenance and sporulation of fungi in culture (Booth, 1971; Singleton *et al.*, 1992), although a large proportion of currently employed protocols are primitive in their design.

Some species of fungi in soil will not grow at all in culture, either because suitable nutrients are not available, or because specific germination and/or growth stimuli are absent. Fungi in this category are usually closely associated with specific plants or animals, and are either present in the soil as survival or dissemination propagules, or as hyphae linked to mycorrhizal associations with plant roots. With current technology, such fungi are more satisfactorily sampled in conjunction with their hosts. Techniques involving culture of soil fungi mostly include antibiotics in the agar, to prevent bacterial contamination. Whereas such a practice is generally advisable, it is not clear whether all fungi can grow in the presence of these substances. There is an enormous range of agar recipes to choose from when isolating fungi (see e.g. Booth, 1971). The composition is not critical apart from the nutrient content (see below) and it seems that in most cases multiplication of agar types in diversity experiments is of lesser value than altering growth conditions and baiting (Bills and Polishook, 1994). A small selection of recipes for media is given below.

Rapidly growing fungi often overgrow more slowly developing colonies, so that unless the culture process is carefully monitored, many species will never be recovered. Protocols to promote the isolation of rare species include using agar media with very low nutrient content and relatively low incubation temperatures; both these measures slow growth and allow easier observation of colonies in the early stages of formation. Cyclosporin has been used successfully as a general growth inhibitor (Dreyfuss, 1986); it has a contact effect but is not taken into the mycelium, thus strongly reducing colony growth but not killing the fungus itself. Many other growth inhibitors have been used in selective processes; a recent example was detailed by Wildman (1991), who used lithium chloride as a selective inhibitor of *Trichoderma* colonies, allowing the development of slower-growing taxa. Colonies can then be transferred to new agar plates lacking inhibitors to promote sporulation. In a proportion of plates, rapidly growing colonies can be excised using a sterile scalpel or burnt off using a soldering iron, allowing the slower growers to develop without competition.

Some studies have demonstrated clear seasonality in the profiles of soil fungus species recovered (Clarke and Christensen, 1981). Sampling therefore needs to occur on several occasions during the year. More long-term fluctuations are poorly researched, and it is difficult to separate changes in fungal composition of soil samples caused by environmental factors such as climate change from the results of substrate decomposition and seral succession (Gams, 1992).

General Sampling Methods

These have been reviewed by many workers, most recently by Frankland *et al.* (1990), Gams (1992), Singleton *et al.* (1992) and Parkinson (1994). Gams (1992) considers that of the three methods detailed below, the soil-washing technique is most generally effective. The equipment is marginally more difficult to sterilize than for the other methods, but it is quite feasible for even poorly equipped laboratories to carry out research programmes using soil-washing techniques. Bååth (1988) has provided a recent critique of the method.

Dilution plate technique

Soil is dried and crushed through a 2 mm diam sieve. Small subsamples are added to 0.2% (w/v) water agar in a screw-capped bottle, and mixed thoroughly. Quantities of the resulting mixture are diluted with further aliquots of water agar, and the process repeated as necessary. Samples of the final dilution are placed in sterile Petri dishes, and molten agar at about 45°C added. Soil particles are dispersed by circular motion of the Petri dish. The number and degree of dilution stages will be determined by the number of fungal particles recovered from the soil; for optimum recovery the number of colonies on each plate must not be greater than 8-10.

Warcup plate technique

Small soil samples (0.005-0.15 g) are transferred to sterile Petri dishes and crushed in drops of water. Molten but cooled agar is added and the particles dispersed as for the dilution plate technique. This technique is simple and rapid, but gives a less accurate picture of diversity in a large soil volume unless a considerable number of plates are prepared. The growth of bacteria is discouraged due to soil particles being deeply submerged in the agar (Frankland, 1990).

Soil-washing technique

Soil samples are agitated in water, either by shaking or passing air through the system. The resulting mixture is then passed through a series of sieves with decreasing mesh size, and washed through with further quantities of water. Mineral and organic particles from one or other of the smaller sieves are then plated out onto agar. Appropriate equipment is illustrated by Bissett and Widden (1972), Gams (1992) and Kirby *et al.* (1990), and detailed also by Bills and Polishook (1994) in a slightly modified form. The serial sieves needed can be made up from items in standard laboratory equipment catalogues. Appropriate particle sizes have been calculated as between 100 and 200µm diam; as Bååth (1988) noted, the ideal size permits a 1:1 colony: particle ratio.

Selective Sampling Methods

There is a wide, but scattered literature of methods for increasing the likelihood of sampling particular taxonomic or ecological groups of fungi, designed either to encourage growth and sporulation of fungi by providing specific nutrients or to encourage growth and sporulation of fungi by providing specific nutrients or physical substrata, or by including compounds toxic to other groups in isolation media. Booth (1971) has provided an extensive list of general and selective media. Some of these methods are well-established and generally applicable, but, as soil characteristics vary enormously throughout the world, selective systems may only operate satisfactorily in the ecosystem for which they were designed (Singleton *et al.*, 1992). Most have not been tested extensively to explore the diversity of species for which media are selective within the claimed target groups. It is not always clear from published literature whether recommended media are truly selective, or simply good for growing particular groups of fungi. Some techniques appear dangerously close to alchemy, and their efficacy is probably more due to chance than to good science.

Selective Media

Using the techniques described above in conjunction with selective media will encourage growth of fungal groups with specific nutritional requirements at the expense of generalists. In order to operate successfully, most selective media will contain low levels of nutrients, in order to discriminate against rapidly growing non-specialist taxa.

Basidiomycete-selective media

Media containing benomyl and phenolic compounds such as lignin derivatives are selective for many groups of soil basidiomycetes (Edgington *et al.*, 1971; Hunt and Cobb, 1971; Watling, 1971; Singleton *et al.*, 1992). Various modifications are used as selective media for *Rhizoctonia* and its relatives. These techniques are selective due to the natural nutritional strategies of wood-associated basidiomycetes, and do not necessarily inhibit the growth of other fungi from similar microhabitats (e.g. Barnard and Cannon, 1987). Little research has been carried out on the variety of basidiomycetes whose growth and isolation is promoted by such methods.

Actidione-containing media

Actidione suppresses respiration in a wide range of fungal species (Anderson and Domsch, 1975), and its use in agar has been used in selection for dermatophyte fungi (*Ascomycota*, *Onygenales*). Some other fungal groups are also resistant, including the yeast genera *Brettanomyces* and *Dekkera* (van der Walt and Yarrow, 1984).

Selective media for specific genera

A large number of media has been designed to aid recovery of particular groups from soil. These include peptone-PCNB agar (Nelson *et al*., 1983) for *Fusarium* species, a selective medium for *Gaeumannomyces* and *Magnaporthe* (Juhnke *et al*., 1984), and dichloran media for *Penicillium* species (Hocking and Pitt, 1980). Media containing cycloheximide are used to differentiate between the genera *Ceratocystis* and *Ophiostoma* (Harrington, 1981). Such techniques would probably not increase the total species count significantly in a carefully conducted inventory, but instead offer methods for promoting rapid recovery of particular groups.

Baiting Techniques

A wide variety of baiting techniques has been developed for sampling particular fungal groups. These are either buried in soil and subsequently recovered and examined directly or plated out on agar, or may be included in culture media to encourage growth and sporulation of propagules from soil samples. Baits are usually designed to provide nutrients which mimic the natural resources utilized by fungi. Other systems provide physical substrata to allow fruit body development where growth in agar culture alone is insufficient for sporulation. Many of these added substrata also act as nutrient sources.

Cellulose

Cellulose may be added to soil by burying 'Cellophane' sheet (washed to remove plasticizers) attached to glass slides as carriers in order to promote development of cellulolytic species (Tribe, 1957; Gams, 1960; Barron, 1971). These can either be examined directly, or plated onto antibiotic media. The addition of filter paper to agar cultures in order to promote fruit body production is also well established (Booth, 1971).

Plant tissues

Living plant parts are used routinely by some plant pathologists as bait for isolation of pathogenic taxa. These include apples (for various fungi including *Cylindrocarpon* and *Colletotrichum*), carrot discs (for *Thielaviopsis*) and *Pelargonium* leaves for *Cylindrocladium* species (Barron, 1971; Singleton *et al*., 1992). These techniques apparently work well for wide-spectrum necrotrophs, but may not be particularly appropriate for inventory work.

Wood pieces

Pieces of wood and bark (normally chips or sawdust) are incorporated into agar to increase the likelihood of isolating a wide range of fungi, ranging from wood-rotting basidiomycetes to saprobic wood-surface microfungi. The wood pieces are normally sterilized and mixed into molten agar, or boiled in water and the liquid incorporated into the culture medium. Transverse slices of newly cut wood are also used for isolation of basidiomycetes such as *Heterobasidion* (Singleton *et al.*, 1992), placing them in Petri dishes and adding aliquots of soil suspensions onto their surface.

Animals

Fungal parasites of nematodes and rotifers can occasionally be sampled using soil dilution or washing methods, but their recovery can be promoted by using suitable animals as bait in weak soil slurries or standard culture media. Methods are described by Barron (1982) and Gray (1984). Nematode cysts containing fungi can also be separated from soil samples using washing techniques (Tribe, 1977; Carris *et al.*, 1989).

Dung

Many fungi which are adapted for life on dung are recovered from soil samples. Their isolation may be encouraged by the incorporation of sterile dung pellets into standard media, or adding filtrates from boiled and cooled dung/water mixtures (Booth, 1971). The spores of many such fungi will not germinate unless a heat or chemical shock is applied, mimicking passage through an animal gut (Dix and Webster, 1995). Mollusc gut enzymes facilitate rapid germination of the sexual spores of some zoosporic 'fungi' (see Chapter 10), and the technique may also promote spore germination for some true fungi.

Keratin

Baits incorporating sterilized hair are commonly used for the selective isolation of keratinophilic fungi (principally members of the *Ascomycota*: *Onygenales*). Petri dishes are half-filled with moistened soil, and small pieces of the bait scattered over the soil surface (Stockdale, 1971). Hair from a range of animal species is recommended to produce the widest variety of such fungi from soil samples. Specialized media are often necessary to maintain growth and sporulation.

Physical and Chemical Parameters

Temperature

A small proportion of fungal species is adapted to survival and growth at high or low temperatures (Carreiro and Koske, 1992; Mouchacca, 1995; Petrini *et al.*, 1992). Incubation at a range of temperatures from around 5 to 40°C is advisable to ensure that such fungi are sampled. Although ambient temperatures in the ecosystem sampled should be considered when setting incubation temperature protocols, thermophilic fungi do occur in soils which would not normally reach extreme temperatures (Apinis, 1963).

Water relations

Some fungi are adapted for growth under extreme osmotic stress (Hocking and Pitt, 1980). Although most species do not commonly occur in soil, they may be selectively isolated using media with high water tension, usually achieved by adding sugars to the medium.

Ethanol pasteurization

This process kills thin-walled hyphae and spores, allowing development of slower-growing pigmented species and germination of survival spores, sclerotia etc. Soil samples are agitated with 60% ethanol for a short time, before being mixed into molten agar and allowed to cool. Barron (1971) describes a similar process using steam for partial sterilization of soil samples, and heat shocks produced by standard techniques or microwaves will probably serve the same purpose. Media incorporating rose bengal have been commonly used for similar reasons in the past as this substance inhibits growth of fast-growing hyaline species (King *et al.*, 1979), but the substance is toxic to a wide range of fungi (Gams, 1992).

Specialized Protocols

Fungi from various specific microhabitats within the soil present particular problems in diversity assessment, and need special treatment.

Litter fungi

Fungi from decaying leaves and twigs on and in the surface layers of soil are extremely diverse (Bills and Polishook, 1994), and far outnumber true soil species in their variety. Where plant parts are recognizable, direct examination using a dissecting microscope, perhaps after incubation in a damp chamber, will reveal many species. Isolation by plating out fragments of plant material onto agar will result in a wide variety of colonies. In order that rare species are not swamped by rapidly growing species, litter samples are dried and pulverized, then washed and filtered, and particles between 100 and 200 μm in size are re-suspended in sterile water and plated out. The technique is very similar to the soil-washing protocol described above. The particle size adopted by Bills and Polishook (1994) appeared to approach the one fungal propagule per particle ideal advocated by Bååth (1988). The necessity for drying samples before pulverization may affect the viability of some species, and research into this question might be advisable before designing protocols for sampling of waterlogged soils or sediments.

Hypogeous fungi

Hypogeous fungi (i.e. truffles) produce fertile fruiting bodies in soil, and can be identified directly from soil samples (Pegler *et al.*, 1993). These fungi have affinities with various of the major fungal groups, and are often mycorrhizal. They are important elements of the soil biota in some soils, especially in deserts. They may be found through careful search of litter layers of suitable soils, though they are harvested commercially using trained animals. Soil sieving can be effective for recovering sclerotia and large spores such as those of arbuscular mycorrhizal fungi (Barron, 1971), and would certainly result in recovery of truffle fruit bodies.

Root endophytes

Endophytic fungi in roots (as opposed to mycorrhizal taxa) have rarely been sampled. Techniques involve the plating out of fragments cut from surface-sterilized roots, using similar techniques to those used to study wood endophytes (Chapela and Boddy, 1988; Carroll, 1991). Most species are probably more or less dormant within the living tissues, growing saprotrophically once the root has died.

Root pathogens

A wide range of fungi are pathogenic on roots, especially those causing wilt symptoms of aerial plant parts. Prominent examples include *Fusarium* species, and the *Gaeumannomyces/Phialophora* complex associated with grass roots, including many cereal crops. A series of general and specific techniques for isolation of pathogenic soil fungi is given in Singleton *et al.* (1992). Effective sampling of such fungi from undisturbed vegetation, particularly in the tropics, may be ineffective due to poor knowledge of pathogenic root fungi associated with non-crop plants. Immunological techniques (principally ELISA) have been used to detect some pathogenic taxa *in situ* (Frankland *et al.*, 1990), but are of value only when particular taxa are suspected of being present.

Rhizosphere fungi

A range of pathogenic and saprobic fungi occur in close association with plant root surfaces, although they are not mycorrhizal. The number of propagules in the vicinity of roots may be many times that of soil samples without roots (Dix and Webster, 1995). Many will be detected using standard soil-washing or filtering techniques, but methods involving root washing have also proved effective (Parkinson and Thomas, 1965; Persiani and Maggi, 1988).

Freshwater and Marine Sediments

No special techniques are required for studying freshwater sediments. Fungi in marine sediments have been patchily studied. A number of specialized media has been developed for growth of these fungi in culture (Jones, 1971), although not all fungi adapted to marine environments require saline media for healthy growth (Jennings, 1986). The mycobiota of estuarine and marine sediments includes many terrestrial elements, presumably washed into the sea by rivers (Kohlmeyer and Kohlmeyer, 1979; Ueda, 1980; White *et al.*, 1980). However, it is not known what proportion of such fungi are metabolically active (Pugh, 1968).

Salt marsh soils have been investigated by a number of authors, and a very wide range of fungi have been found associated with detritus, especially of grass species. Only a small proportion of these have been recovered by culturing soil samples (Gessner and Kohlmeyer, 1976; Kohlmeyer and Kohlmeyer, 1979). Mangrove ecosystems have been intensively studied, but studies have almost exclusively concentrated on fungi of lignified substrates (Kohlmeyer and Kohlmeyer, 1979). Again there appears little evidence as to the extent of biotic activity of the species recovered. Miller and Whitney (1981) and Molitoris and Schaumann (1986) refer to various fungi in marine sediment samples, but there has been no detailed or intensive research into filamentous fungi from deep-sea sediments.

Molecular Techniques

Molecular methods are now widely used to differentiate between related taxa and to detect individual species in environmental samples (Correll, 1992). Their use in diversity studies is still in its infancy.

Torsvik *et al.* (1990) used a novel molecular technique for study of bacterial diversity in a Norwegian soil sample which involves separation of bacterial DNA from that of other organisms and sequencing the products to provide an estimate of the variety of DNA molecules present. A far higher level of diversity was identified than by using culturing, suggesting that a large proportion of bacterial species present cannot be detected using traditional methods. Similar approaches may be possible with fungi in soil samples, but the problems of scale and interpretation mean that the technique is not yet practical. There are difficulties in equating the taxa identified through studies of sequence variation with those accepted using morphological research, and current technology can only provide estimates of numbers rather than actually identifying more than a tiny proportion.

Limitations of Existing Techniques

Current techniques for measuring the diversity of fungi in soil and sediment samples have some limitations. First among these is that not all species of fungi present in soil will grow in culture using standard techniques. Most of these are mycorrhizal, and can be sampled in association with plant roots (see Chapter 12). Others are also closely associated with other organisms (e.g. as antagonistic or mutualistic biotrophs), and are present either as survival propagules, or by chance due to failure of the dissemination process. In most cases, this last group will most efficiently be sampled along with their associated taxa, and will play an insignificant role in soil ecosystem processes. It is possible that there are non-culturable fungi in the soil which could only be detected using molecular techniques such as those described above, but perhaps this is unlikely due to the size of fungal structures in comparison with those of bacteria.

A second and more major problem is that a significant proportion of those fungi which will grow using standard culture techniques will either not sporulate at all, or not maintain their fertility after subculture. Sporulating structures are necessary for identification in almost all cases using current technology. Not only are non-sporulating fungi very difficult to identify, but the characters of the vegetative mycelium are often hardly sufficient to separate morphotypes. Consequently, estimating numbers of species is difficult.

A third challenge for those sampling fungal diversity in soil is that due to the size of and variation within the fungal kingdom, it is unlikely that a single protocol will prove a satisfactory method of diversity estimation. Using several methods will unavoidably result in some duplication of effort. There is also little information on the reliability of individual methods and the extent to which protocols which are effective for one soil ecosystem can successfully be transferred to others. Poor knowledge of seasonality of most fungi in soil is also a limiting factor, although this may be less relevant for diversity sampling than ecological study, as seasonal effects may largely be quantitive rather than qualitative. There have been attempts to develop universal protocols for the ecological study of soil fungi in the past (International Biological Programme; Frankland, 1990), but this current initiative will do much to ensure that studies of fungal soil diversity can be properly related to one another.

Overcoming Limitations of Techniques

The ability to provide rapid and complete assessments of fungal diversity is still some way off (Rossman, 1994). Rapid progress could be made in some research areas, which would result in clear improvements in inventory technique. These include:

- Proper assessment of current techniques for fungal culture, designed to identify protocols which are complementary but result in minimum overlap of species.
- Consideration of how these protocols should vary with different soil conditions and ecosystems.
- Research into the inclusion of novel compounds into media, with the aim of improving germination, sporulation and maintenance in culture.
- Development of novel methods for characterizing strains and species from non-sporulating cultures.
- Research into non-axenic culture methods.

Integrating Techniques into a General Scheme of Analysis

There is currently little information on the variation in diversity recorded using different isolation and culture techniques, though preliminary assessments by Christensen (1981) and Gams (1992) provide a platform for further study. It will be advisable to include an experimental stage in the development of protocols assessing the results and practicalities of the different available methods for soil analysis on a series of samples from widely differing ecosystems and seasons. The results can then be quantified using complementarity techniques (Colwell and Coddington, 1994). Once this has been completed, we shall be in a satisfactory position to recommend a specific suite of protocols.

Culture Media Formulations

CYCL (Bills and Polishook, 1994)

Malt extract	10 g
Yeast extract	2 g
Agar	20 g
Cyclosporin A	10 mg
Chlorotetracycline	50 mg
Streptomycin sulphate	50 mg
Water	1 l

The last three ingredients are added once the medium has cooled after autoclaving.

DRBC (Bills and Polishook, 1994)

Peptone	5 g
Dextrose	10 g
Potassium phosphate	1 g
Magnesium sulphate	0.5 g
Dichloran	0.002 g
Rose Bengal	0.025 g
Chloramphenicol	0.01 g
Agar	15 g
Water	1 l
Chlorotetracycline	50 mg
Streptomycin sulphate	50 mg

The last two ingredients are added once the medium has cooled after autoclaving.

CMCM (CARBOXYMETHYLCELLULOSE MEDIUM; Bills and Christensen, unpubl.)

Sodium carboxymethylcellulose	10 g
Yeast extract	0.5 g
Ammonium sulphate	1 g
Potassium nitrate	2 g
Potassium phosphate	1 g
Magnesium sulphate	0.5 g
Potassium chloride	0.5 g
Calcium chloride	50 mg
Ferrous sulphate	10 mg
Copper sulphate	10 mg
Manganese sulphate	5 mg
Zinc sulphate	1 mg
Agar	15 g
Water	1 l
Chlorotetracycline	0 mg
Streptomycin sulphate	50 mg

The last two ingredients are added once the medium has cooled after autoclaving. The medium is completely transparent, which aids detection of growing colonies, and permits only very slow growth. Care must be taken to mix the ingredients thoroughly as and after water is added, as sodium carboxymethylcellulose is only sparingly soluble.

MEA (Booth, 1971; note quarter-strength)

Malt extract	5 g
Agar	20 g
Water	1 l
Chlorotetracycline	50 mg
Streptomycin sulphate	50 mg

The last two ingredients are added once the medium has cooled after autoclaving.

PCA (from Booth, 1971; note half-strength)

Potato (washed, peeled and grated)	10 g
Carrot (washed, peeled and grated)	10 g
Agar	20 g
Water	1 l
Chlorotetracycline	50 mg
Streptomycin sulphate	50 mg

The potato and carrot are boiled in the water for one hour, and passed through a fine sieve. The agar is then added, boiled until dissolved and then autoclaved. The last two ingredients are added once the medium has cooled after autoclaving.

WATER AGAR

Agar	20 g
Tap water	1 l
Chlorotetracycline	50 mg
Streptomycin sulphate	50 mg

The last two ingredients are added once the medium has cooled after autoclaving.

Acknowledgements

At a late stage in preparation of this manuscript, I was given access to a draft contribution for a forthcoming publication entitled *Evaluating Diversity of Soil Fungi* by the authors Gerald Bills (Merck Research Laboratories, New Jersey) and Martha Christensen (University of Wyoming). Although the manuscript did not affect the overall structure and content of this chapter materially, I found it an invaluable source of further references.

Bibliography

Anderson, J.P.E. and Domsch, K.H. (1975) Measurements of bacterial and fungal contributions to respiration of selected agricultural and forest soils. *Canadian Journal of Microbiology* 21, 314-322.

Apinis, A.E. (1963) Occurrence of thermophilous microfungi in certain alluvial soils near Nottingham. *Nova Hedwigia* 5, 57-78.

Bååth, E. (1988) A critical examination of the soil washing technique with special reference to the size of the soil particles. *Canadian Journal of Botany* 66, 1566-1569.

Barnard, E.L. and Cannon, P.F. (1987) A new species of *Monascus from* pine tissues in Florida. *Mycologia* 79, 479-484. [Ineffectiveness of basidiomycete-selective techniques.]

Barron, G.L. (1968) *The Genera of Hyphomycetes from Soil.* Williams and Wilkins, Baltimore, 364 pp. [Identification manual.]

Barron, G.L. (1971) Soil fungi. In: Booth, C. (ed) *Methods in Microbiology*, vol. 4. Academic Press, London and New York, pp. 405-427. [Valuable general text.]

Barron, G.L. (1982) Nematode-destroying fungi. In: Burns, R.G. and Slater, J.H. (eds) *Experimental Microbiology*. Blackwell, Oxford, pp. 533-552. [Includes protocols for sampling this specialized group.]

Bills, G.F. and Polishook, J.D. (1994) Abundance and diversity of microfungi in leaf litter of a lowland rain forest in Costa Rica. *Mycologia* 86, 187-198. [Techniques for assessment of litter diversity; emphasizes enormous extent of fungal diversity in such habitats.]

Bissett, J. and Widden, P. (1972) An automatic, multichamber soil-washing apparatus for removing fungal spores from soil. *Canadian Journal of Microbiology* 18, 1399-1404.

Booth, C. (1971) Fungal culture media. In: Booth, C. (ed) *Methods in Microbiology,* vol. 4. Academic Press, London and New York, pp. 49-94. [Invaluable general text including a wide range of media recipes.]

Carreiro, M.M. and Koske, R.E. (1992) Room temperature isolations can bias against selection of low temperature microfungi in temperate forest soils. *Mycologia* 84, 886-900.

Carris, L.M., Glawe, D.A., Smyth, C.A. and Edwards, D.I. (1989) Fungi associated with populations of *Heterodera glycines* in two Illinois soybean fields. *Mycologia* 81, 66-75. [Protocol for recovery of fungi from nematode cysts.]

Carroll, G.C. (1991) Fungal associates of woody plants as insect antagonists in leaves and stems. In: Barbosa, P., Krischik, V.A. and Jones, C.G. (eds) *Microbial Mediation of Plant-Herbivore Interactions.* Wiley, New York, pp. 253-271. [Protocols for isolation of wood and root endophytes.]

Chapela, I.H. and Boddy, L. (1988) Fungal colonization of attached beech branches. II. Spatial and temporal organization of communities arising from latent invaders in bark and functional sapwood under different moisture regimes. *New Phytologist* 110, 47-57. [Protocols for wood (and bark) endophytes.]

Christensen, M. (1981) Species diversity and dominance in fungal communities. In: Wicklow, D.T. and Carroll, G.C. (eds) *The Fungal Community. Its Organization and Role in the Ecosystem.* Dekker, New York, pp. 201-232.

Christensen, M. (1989) A view of fungal ecology. *Mycologia* 81, 1-19 [Comprehensive text emphasizing the role of fungi in ecosystems, including soil.]

Clarke, D.C. and Christensen, M. (1981) The soil microfungal community of a South Dakota grassland. *Canadian Journal of Botany* 59, 1950-1960.

Colwell, R.K. and Coddington, J.A. (1994) Estimating terrestrial biodiversity through extrapolation. *Philosophical Transactions of the Royal Society of London Biological Sciences*, series B, 345, 101-118.

Correll, J.C. (1992) Genetic, biochemical, and molecular techniques for the identification and detection of soilborne plant-pathogenic fungi. In: Singleton, L.L. Mihail, J.D. and Rush, C.M. (eds) *Methods for Research on Soilborne Phytopathogenic Fungi*. APS Press, St Paul, pp. 7-16.

Dix, N.J. and Webster, J. (1995) *Fungal Ecology*. Chapman and Hall, London, 549 pp.

Domsch, K.H., Gams, W. and Anderson, T.-H. (1993) *Compendium of Soil Fungi*, rev. edn, 2 vols. Academic Press, London. [The best text for identification of soil fungi, although despite its wide scope it is by no means comprehensive.]

Dreyfuss, M.M. (1986) Neue Erkenntnisse aus einem pharmakologischen Pilz-Screening. *Sydowia* 39, 22-36. [Protocol using cyclosporin as a growth inhibitor.]

Edgington, L.V., Khew, K.L. and Barron, G.L. (1971) Fungitoxic spectrum of benzimidazole compounds. *Phytopathology* 61, 42-44. [Selective media.]

Frankland, J.C. (1990) Ecological methods of observing and quantifying soil fungi. *Transactions of the Mycological Society of Japan* 31, 89-101. [General review paper.]

Frankland, J.C., Dighton, J. and Boddy, L. (1990) Methods for studying fungi in soil and forest litter. *Methods in Microbiology* 22, 343-404.

Gams, W. (1960) Studium zellulolytischer Bodenpilze mit Hilfe der Zellophanstreifen-Methode und mit Carboxylmethyl-Zellulose. *Sydowia* 14, 295-307. [Use of baits for cellulolytic fungi, carboxymethylcellulose agar.]

Gams, W. (1992) The analysis of communities of saprophytic microfungi with special reference to soil fungi. In: Winterhoff, W. (ed.) *Fungi in Vegetation Science*. Dordrecht, Netherlands Kluwer, pp. 183-223. [Excellent recent text comparing sampling techniques.]

Gams, W. and Domsch, K.H. (1969) The spatial and seasonal distribution of microscopic fungi in arable soils. *Transactions of the British Mycological Society* 52, 301-308.

Gessner, R.V. and Kohlmeyer, J. (1976) Geographical distribution and taxonomy of fungi from marsh *Spartina*. *Canadian Journal of Botany* 54, 2023-2037.

Gray, N.F. (1984) Ecology of nematophagous fungi: methods of collection, isolation and maintenance of predatory and endoparasitic fungi. *Mycopathologia* 86, 143-153.

Hall, G. (1987) Sterile fungi from roots in winter wheat. *Transactions of the British Mycological Society* 89, 447-456.

Harrington, T.C. (1981) Cycloheximide sensitivity as a taxonomic character in *Ceratocystis. Mycologia* 72, 1123-1129. [Selective media.]

Hawksworth, D.L. (1991) The fungal dimension of biodiversity: magnitude, significance, and conservation. *Mycological Research* 95, 641-655.

Hawksworth, D.L., Kirk, P.M., Sutton, B.C. and Pegler, D.N. (eds) (1995) *Ainsworth and Bisby's Dictionary of the Fungi*, 8th edn. CAB INTERNATIONAL, Wallingford. [Essential basic reference work.]

Hocking, A.D. and Pitt, J.I. (1980) Dichloran-glycerol medium for enumeration of xerophilic fungi from low-moisture foods. *Applied and Environmental Microbiology* 39, 488-492. [Methods for isolation of osmotolerant fungi.]

Hunt, R.S. and Cobb, F.W. (1971) Selective medium for the isolation of wood-rotting basidiomycetes. *Canadian Journal of Botany* 49, 2064-2065.

Jennings, D.H. (1986) Fungal growth in the sea. In: Moss, S.T. (ed.) *Biology of Marine Fungi*. Cambridge University Press, Cambridge, pp. 1-18.

Jones, E.B.G. (1971) Aquatic fungi. In: Booth, C. (ed.) *Methods in Microbiology*, vol. 4. Academic Press, London and New York, pp. 335-365.

Juhnke, M.E., Mathre, D.E. and Sands, D.C. (1984) A selective medium for *Gaeumannomyces graminis* var. *tritici*. *Plant Disease* 68, 233-236.

Keyworth, W.G. (1951) A Petri-dish moist chamber. *Transactions of the British Mycological Society* 31, 291-292.

King, A.D., Hocking, A.D. and Pitt, J.I. (1979) Dichloran-rose bengal medium for the enumeration and isolation of molds from foods. *Applied and Environmental Microbiology* 37, 959-964.

Kirby, J.J.H., Webster, J. and Baker, J.H. (1990) A particle plating method for analysis of fungal community composition and structure. *Mycological Research* 94, 621-626.

Kjøller, A. and Struwe, S. (1982) Microfungi in ecosystems: fungal occurrence and activity in litter and soil. *Oikos* 39, 391-422.

Kohlmeyer, J. and Kohlmeyer, E. (1979) *Marine Mycology. The Higher Fungi*. Academic Press, New York, 690 pp. [Good book concentrating on enumeration and classification but with some ecological information.]

Lodge, D.J. (1993) Nutrient cycling by fungi in wet tropical forests. In: Isaac, S., Frankland, J.C., Watling, R. and Whalley, A.J.S. (eds) *Aspects of Tropical Mycology*. Cambridge University Press, Cambridge, pp. 37-57.

Miller, J.D. and Whitney, N.J. (1981) Fungi from the Bay of Fundy III. Geofungi in the marine environment. *Marine Biology* 65, 61-68.

Molitoris, H.P. and Schaumann, K. (1986) Physiology of marine fungi: a screening programme for growth and enzyme production. In: Moss, S.T. (ed.) *The Biology of Marine Fungi*. Cambridge University Press, Cambridge, pp. 35-47.

Moubasher, A.H. (1993) *Soil Fungi in Qatar and Other Arab Countries*. University of Qatar, Doha, 568 pp. [Identification manual.]

Mouchacca, J. (1995) Thermophilic fungi in desert soils - a neglected extreme environment. In: Allsopp, D., Colwell, R.R. and Hawksworth, D.L. (eds)

Microbial Diversity and Ecosystem Functioning. CAB INTERNATIONAL, Wallingford, pp. 265-288.

Nelson, P.E., Toussoun, T.A. and Marasas, W.F.O. (1983) *Fusarium Species: An Illustrated Manual for Identification*. Pennsylvania State University Press, Pennsylvania. [Selective media.]

Parkinson, D. (1994) Filamentous fungi. In: Weaver, R.W., Angle, S., Bottomley, P., Bezdicek D. and Smith, S. (eds) *Methods of Soil Analysis. Part 2. Microbiological and Biochemical Properties*. SSSA Book Series no. 5, Soil Science Society of America, Madison, pp. 329-350. [General techniques.]

Parkinson, D. and Thomas, A. (1965) A comparison of methods for the isolation of fungi from rhizospheres. *Canadian Journal of Microbiology* 11, 1001-1007.

Parkinson, D., Gray, T.R.G. and Williams, S.T. (eds) (1971) *Methods for Studying the Ecology of Soil Micro-Organisms*. IBP Handbook No. *19*. Blackwell, Oxford, 116 pp. [An old general text but many techniques still widely used.]

Pegler, D.N., Spooner, B.M. and Young, T.W.K. (1993) *British Truffles. A Revision of British Hypogeous Fungi*. Royal Botanic Gardens, Kew, 216 pp.

Persiani, A.M. and Maggi, O. (1988) Fungal communities in the rhizosphere of *Coffea arabica* L. in Mexico. *Micologia Italiana* 2, 21-37. [Root-washing protocol.]

Petrini, O., Petrini, L.E. and Dreyfuss, M.M. (1992) Psychrophilic deuteromycetes from alpine habitats. *Mycologia Helvetica* 5, 9-20.

Pugh, G.J.F. (1968) A study of fungi in the rhizosphere and on the root surfaces of plants growing in primitive soils. In: Phillipson, J. (ed) *Methods of Study in Soil Ecology*. UNESCO, Paris, pp. 159-164. [Metabolic activity of fungi in marine sediments.]

Rambelli, A., Persiani, A.M., Maggi, O., Lunghini, D., Onofri, S., Riess, S., Dowgiallo, G. and Puppi, G. (1983) *Comparative Studies on Microfungi in Tropical Ecosystems*. UNESCO, Rome.

Rossman, A.Y. (1994) A strategy for an all-taxa inventory of fungal biodiversity. In: Peng, C.-I., and Chou, C.-H. (eds) *Biodiversity and Terrestrial Ecosystems*. Academia Sinica, Taipei, pp. 169-194. [Fungal diversity.]

Singleton, L.L., Mihail, J.D. and Rush, C.M. (eds) (1992) *Methods for Research on Soilborne Phytopathogenic Fungi*. APS Press, St Paul.

States, J.S. (1978) Soil fungi of cold desert plant communities in northern Arizona and southern Utah. *Journal of the Arizona-Nevada Academy of Science* 13, 13-17. [Fungal diversity in soil.]

Stockdale, P.M. (1971) Fungi pathogenic for man and animals I. Diseases of the keratinized tissues. In: Booth, C. (ed.) *Methods in Microbiology*, vol. 4. Academic Press, London and New York, pp. 429-460. [Selective media for keratinophilous fungi.]

Torsvik, V., Goksøyr, J. and Daae, F.L. (1990) High diversity in DNA of soil bacteria. *Applied and Environmental Microbiology* 56, 782-787.

Tribe, H.T. (1957) Ecology of micro-organisms in soils as observed during their development upon buried cellulose film. In: Williams, R.E.O. and Spicer, C.C. (eds) *Microbial Ecology*, pp. 287-298. [Methods for selective isolation of cellulolytic taxa.]

Tribe, H.T. (1977) Pathology of cyst nematodes. *Biological Reviews* 52, 477-507. [Fungi associated with nematode cysts.]

Ueda, S. (1980) A mycofloral study on brackish water sediments in Nagasaki, Japan. *Transactions of the Mycological Society of Japan* 21, 103-112.

van der Walt, J.P. and Yarrow, D. (1984) Methods for isolation, maintenance, classification and identification of yeasts. In: Kreger-van Rij, N.J.W. (ed.) *The Yeasts. A Taxonomic Study*, edn 3. Elsevier, Amsterdam, pp. 45-104.

Wainwright, M. (1988) Metabolic diversity of fungi in relation to growth and mineral cycling in soil - a review. *Transactions of the British Mycological Society* 90, 159-170.

Warcup, J.H. (1955) Isolation of fungi from hyphae present in soil. *Nature* 175, 953.

Watling, R. (1971) *Basidiomycetes. Homobasidiomycetidae*. In: Booth, C. (ed.) *Microbiological Methods*, vol. 4. Academic Press, London and New York, pp. 219-236. [Protocols for selective isolation.]

White, D.C., Bobbie, R.J., Nickels, J.S., Fazio, S.D. and Davis, W.M. (1980) Nonselective biochemical methods for the determination of fungal mass and community structure in estuarine detrital microflora. *Botanica Marina* 23, 239-250.

Wildman, H.G. (1991) Lithium chloride as a selective inhibitor of *Trichoderma* species on soil isolation plates. *Mycological Research* 95, 1364-1368.

Arbuscular Mycorrhizas 12

JUSTIN P. CLAPP, ALASTAIR H. FITTER
AND JAMES W. MERRYWEATHER

Department of Biology, University of York, York YO1 5DD, UK.

Ecological Significance of Arbuscular Mycorrhizas

Mycorrhizal fungi are ubiquitous root symbionts. Around 90% of plant species probably normally form mycorrhizas, but the term covers a wide range of distinct symbiotic associations. The main types are arbuscular mycorrhizas (AM: sometimes called vesicular-arbuscular mycorrhizas or VAM), ectomycorrhizas (EM or ECM), ericoid and orchid mycorrhizas. The last two are found only with a taxonomically limited set of plant hosts. There are a number of other less well-defined or understood types. Ectomycorrhizas and AM are the most significant ecologically; AM probably occur with two-thirds of all plant species (Trappe, 1987). Although mycorrhizas probably originally evolved as phosphate-gathering systems in Silurian or Devonian land plants, their modern functions are more diverse (Table 12.1). Nevertheless, the most important ecosystem function of mycorrhizal fungi is in nutrient cycling; other functions are more manifest at the level of community structure.

Techniques for Studying Arbuscular Mycorrhizas

This section concentrates exclusively on AM since these are by far the most widespread type and, because they are obligate symbionts, they offer the most difficult technical problems.

The Systematics of Arbuscular Mycorrhizal Fungi

We recommend that researchers who require to identify and culture mycorrhizal fungi attend one of the courses which occur from time to time, frequently at

 Methods for the Examination of Organismal Diversity in Soils and Sediments (ed. G.S. Hall).

Table 12.1. Characteristics of the four main types of mycorrhizal association.

Type of mycorrhiza	Plants	Fungus	Fungal carbon source	Plant benefits: uptake of P	uptake of N	uptake of water	protection from pathogens	protection from grazing
Arbuscular mycorrhiza	Moist, except some families (e.g. *Brassicaceae*) and those below	*Glomales* (*Zygomycota*) *Glomus* *Acaulospora*, *Scutellospora*, *Gigaspore*, *Entrophospora*	Entirely from plant	Bypasses diffusion pathways in soil	As P (NH_4^+ only)	Probably by improved root-soil contact	Reduced colonization by e.g. *Fusarium*	Leaves are less palatable
Ecto-mycorrhiza or Sheathing mycorrhiza	Woody plants, especially in boreal and cool-temperate regions, some tropical systems (e.g. in *Caesal-piniaceae* *Diptero-carpaceae*)	Mainly *Basidiomycota*; some *Ascomycota*	Mainly from plant	As above	Some fungi can obtain N from organic sources	May have large diameter hyphal channels, water transport possible	Sheath acts as physical barrier	Uncertain
Ericoid mycorrhiza	*Ericales*	*Ascomycota*	Mainly from plant	Probably as for N but less important	Obtain N from organic sources	Unknown	Unknown	Unknown
Orchid mycorrhiza	*Orchidaceae*	*Basidiomycota* (e.g. *Rhizoctonia*) and related mitosporic fungi	From soil or other plants	Yes	Yes	Unknown	Probably	Unknown

mycorrhiza conferences. Arbuscular mycorrhizal fungi are currently considered all to be members of the order *Glomales*, classified thus:

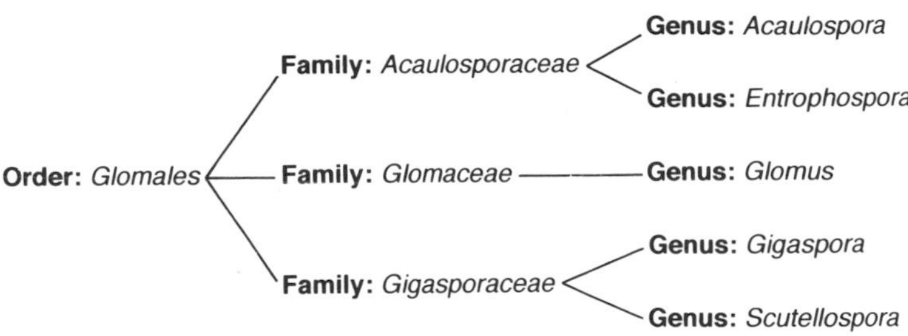

The taxonomy of the *Glomales* is relatively new and, therefore, incomplete. Not only are there probably many taxa yet to be described, but molecular techniques are revealing that the group may be much more diverse than was previously thought. It has proved impossible to classify arbuscular mycorrhizas using features of fungi within colonized roots, although some features may indicate to which genus an endophyte may belong. Therefore, classification is normally based upon the morphological characters of spores, and association with the morphology of their mycorrhizas is only possible in pure culture. Because the taxonomy of the *Glomales* is, relatively speaking, in its infancy, there is no satisfactory key to Glomalean spores.

Both INVAM (International Collection of Vesicular-Arbuscular Mycorrhizal Fungi) and BEG (Banque Européenne des Glomales) are pleased to receive material if it is presented in an acceptable form. Generally speaking, the most useful specimens are living pure cultures from which the recipients may produce their own cultures, as well as herbarium material in the form of microscope slides and spores in preservative. Thc last is rarely wholly satisfactory because microscopic features of taxonomic importance change gradually in all preservatives used to date. Extreme caution should be exercised before naming a spore, spore assemblage or culture. We recommend that the taxonomists at INVAM or BEG be consulted whenever an arbuscular mycorrhiza is to be described. (See Useful Addresses.)

Methods for the Extraction of Glomalean Spores from Soils

Soils generally contain a large proportion of organic matter, sufficient to require a particular method of separating spores from the bulk of the soil from which they are being sampled (sucrose flotation). The quick method detailed below is often highly successfully applied to carefully designed pot culture media.

Sucrose flotation

Collect soil from around host plant roots.

- Thoroughly wet a known weight of fresh soil (e.g. 100 g).
- Weigh a subsample of soil, dry to constant weight at 70°C, and reweigh for calculation of number of spores per unit dry soil.
- Pass soil suspension through a 710 mm Endecotts sieve, removing stones and roots. Strain the soil suspension through a fine sieve (32 mm) and transfer the solid matter collected to four 50 ml centrifuge tubes.
- Add water, balance the tubes and resuspend the soil sample. Centrifuge at 1800 rpm for 5 min.
- Discard the supernatant, which contains floating organic material, including dead spores. Resuspend pellets in sucrose solution (440 g l^{-1}). Carefully balance the tubes.
- Centrifuge up to 1800 rpm and brake immediately.
- Rapidly sieve the supernatant (32 mm) and wash thoroughly (at least 1 min) to replace the sucrose and alleviate osmotic stress on spores.
- Discard the pellets left in the centrifuge tubes.
- Carefully wash all of the solid material from the sieve into a Petri plate marked with a grid to facilitate spore counting and collection under a dissecting microscope.
- Alternatively (if there is little matter other than spores in the sample) transfer to a filter paper via a Büchner funnel.
- Lift and transfer spores with ultra-fine forceps. This is perfectly practical with spores of as little as 50 mm diam.
- If spores are to be used for culture attempts, allow them to recover from the stress of extraction in axenic water for 24 h at 4°C.

Rapid spore extraction

For soils with low organic matter content.

- Take a soil sample and place it in a beaker.
- Vigorously add more than twice as much water and allow to soak for a few minutes.
- Swirl, allow heavy particles to settle and quickly pour off the resulting supernatant into a 32 mm sieve.
- Repeat about 10 times.
- Check the remaining soil for spores. There should be none.
- Carefully wash all of the solid material from the sieve into a Petri plate marked with a grid to facilitate spore counting and collection under a dissecting microscope.

Making Microscope Slides of Glomalean Spores

Microscope slides are essential both for the investigation of spore morphology and as herbarium specimens to represent preserved spores, notes and cultures. A single slide should carry two sets of spores, one mounted in PVLG and another in PVLG/Melzer's reagent.

- Extract spores according to methods given above and store them in water in a small watch glass.
- Take a microscope slide (Chance Propper: 76 × 26 mm, 1.0-1.2 mm thick, with ground glass panels on both sides at one end for pencil labelling) and place on it two small drops of PVLG.
- To one drop of PVLG add a small drop of Melzer's reagent and mix thoroughly with a mounted needle. (If PVLG and Melzer's reagent have been premixed, 1:1 colours may be compared slide to slide).
- Using ultra-fine forceps, transfer spores from watch glass to drops on slide in equal numbers. If available, use large numbers of spores (>10-50) so that all aspects may be seen.
- With fine forceps, gently lower a cover slip (Chance Propper No. 0, 13 mm diameter) onto the drops of mountant. **Do not crush the spores**. Their gross morphology and dimensions may be observed and measured, but little else will be seen.
- Then, carefully apply pressure to the cover slip above the spores with a mounted needle (clean off any stray PVLG) to
  i) crush them gently so that they pop open, then
  iii) crush them further so that the inner spore membranes are disrupted and emerge. Melzer's reaction, if positive, is often rapid during this stage. **Take care not to break the cover slip.**

Experience will indicate what are the most useful treatments you can apply to spores. It is essential to reveal every aspect of structure possible, whilst not destroying the spores.

- Label the slides fully and let them stand for approximately 24 h, during which the spores will clear.
- Place in oven or on slide-warming hot-plate at about 60°C for a further 24 h. The mountant will harden, and the slide become permanent. Excess mountant may be wiped away later with a little water.
- Observe, note, draw, photograph and record all details of spore structure.
- A high-quality compound microscope is an essential tool in determining the internal detail of spores. When bright-field illumination is used, the condenser diaphragm should be open, otherwise the presence of features which do not exist (e.g. layered walls, additional fine walls) may be inferred. Nomarski interference illumination is a useful addition.

- Store safely and, if they are to be used in published descriptions, lodge copies with a relevant, reputable reference collection.

Mountants and stains

PVLG: POLYVYNYL ALCOHOL-LACTO-GLYCEROL	
Polyvinyl alcohol*	1.66 g
Water	10 ml
Lactic acid	10 ml
Glycerin	1 ml

* (99-100% hydrolysed, 24-32 centipoise viscosity)

MELZER'S REAGENT	
Potassium iodide	1.5 g
Iodine	0.5 g
Chloral hydrate (caution)	100 g
Water	22 ml

Preservation of Glomalean Spores

1. Freeze with liquid nitrogen and store in a freezer at -80°C.
2. Freeze-dry.
3. Preserve in liquid, e.g. 5% (v/v) formaldehyde, 3.5% (v/v) glyceraldehyde, or 0.025% (v/v) sodium azide solution, which is currently considered to be the best method, but is also the most hazardous. Lactophenol and formaldehyde acetic alcohol (FAA), used in the past, cause unacceptable damage to glomalean spores.

Isolation and Culture of Glomalean Fungi

The AM fungi of the order *Glomales* are obligately symbiotic and cannot be cultured without a host plant. An arbuscular mycorrhizal culture may be produced from a number of fungal inocula presented to a suitable host plant: soil hyphae, colonized roots and spores. Such inocula, individually or variously combined, may be known (INVAM) as 'germ plasm'. The most desirable AM culture contains a single fungal taxon, achieved by inoculation with a single spore. Such a culture is usually known as an 'isolate', which can be given an identification code and, perhaps, a conventional scientific binomial. Single spore isolates are usually difficult to produce from some field-gathered spores, which may germinate soon after maturing in the wild, and which are disrupted by extraction. Alternative methods of obtaining viable spores are required. 'Trap' culture is widely used to establish mycorrhizal fungi in a suitable host. Sporulation is encouraged and viable spores collected.

Trap culture

Suitablc host plants vary according to geographical location, climate and the preference of the researcher. Select a species which does not produce too much vegetation, will form mycorrhizas with a wide range of glomalean species and is easy to maintain in your locality (Morton, 1995). Single-taxon cultures can sometimes be obtained from single spores extracted from field-collected soil. Success may be limited and variable. Mixed species may be trapped in three ways.

Living plant roots
Plant a mycorrhizal plant in a sterilized sandy potting medium and surround with seedlings (or seeds) of an acceptor species. Remove acceptor plants when colonized with mixed species, and re-pot singly.

Excised mycorrhizal roots
Plant an acceptor species in a sterilized potting medium within which roots have been incorporated.

Mycelial fragments, roots, spores, etc.
Mix soil from around mycorrhizal plants with sterilized sand (1:10 dilution). Plant an acceptor host and await development of mycorrhizas.

When roots are well colonized, small numbers of spores may be extracted for sub-culturing. However, full spore production may be encouraged by removing the aerial parts of the host plant and drying the soil until either almost or totally dry. The root ball is refrigerated at 4°C for several weeks, during which time spore dormancy should be broken and spores rendered suitable for subculturing in large numbers.

Single-spore cultures

Having obtained a supply of viable spores it is necessary to prepare single-taxon isolates. There are several ways of achieving this, but replication is essential to ensure success. Host seedlings may be grown to an appropriate size in sterile sand. If grown thus for too long, however, they will lose vigour.

- Take a host plant seedling and wash sand from the roots under a running tap. The roots will tend to plait together, which is advantageous. Remove the root tip to encourage branching. Under a dissecting microscope, place a single spore within the tangled roots and carefully plant in a suitable potting medium (see below).
- Viable spores may be pregerminated to help ensure success (Brundrett *et al.*, 1994). This, however, may not apply to all taxa, but *Gigaspora* and some

species of *Scutellospora* do seem to respond well to this method. Fresh spores are placed on small pieces of cellulose filter membrane, dialysis tubing or similar, on a bed of sterile moist sand in a Petri plate. Daily observation with a dissecting microscope will determine when germination has occurred. Successfully germinated spores are carefully applied to host roots as above.

- Spores may be germinated to colonize seedling roots or excised (sometimes transformed) roots of an acceptor species growing on a minimal nutrient agar (e.g. White's) in Petri plates. Tomato has proved to be a useful host in this method.

AM culture substrates and conditions

Many culture substrates have been devised by laboratories throughout the world. Success depends on suitability for the selected host species as well as healthy growth of the fungal symbiont. It is important to consider convenience of handling when devising a new substrate, for if it contains a lot of light material, e.g. organic matter, spore extraction will not be simple. If most of the substrate will sink rapidly in water, then flushing the spores out in aqueous suspension will be simple and rapid.

The substrate must also provide the conditions and nutrients essential for the growth of both host and mycosymbiont. The pH must be appropriate. Growth conditions may be judged from the site of origin of the mycorrhiza to be cultured. Most nutrients must be adequately provided, but mycorrhizal establishment is frequently enhanced by a low phosphate, so this nutrient should be limited.

Sterile washed sand is a good base for AM culture. Various clay aggregates have been used, some of which provide an acidic environment, others alkaline. The nutrient supply can be controlled if added in solution (e.g. Hoagland's, Rorison's). If additional nutrient is to be incorporated, a slow-release formulation is preferable, especially one in which phosphate is not immediately available. Rock phosphate and bone meal fulfil this requirement.

Open pots are likely to become infected with unwanted AM fungi (some workers have found that a particular isolate may be extremely invasive), collembola and fungus gnats (which feed on AM fungi), root-feeding fly larvae and fungal leaf pathogens (e.g. powdery mildews). Walker and Vestberg (1994) devised a method of isolating individual cultures in tissue culture bags which permits the maintenance of large numbers of cultures in a small space whilst avoiding contamination. However, cultures in bags must be kept cool, for internal temperatures can rise dramatically in hot weather. This method is probably only suitable for glasshouses in temperate climates and controlled-environment cabinets.

A most important condition for the initiation and maintenance of AM cultures is light. Since the host provides up to 20% of carbon it produces to its fungal partner, it is essential to ensure that photosynthesis is maximized by supplying powerful illumination

Staining Arbuscular Mycorrhizas

Roots are frequently opaque and pigmented, and it is essential to experiment with methods of clearing, bleaching and staining before proceeding with routine staining. The various stains used in mycorrhiza research have subtly different properties and some, such as aniline blue and chorlazol black E, stain all fungal tissue very darkly, often too intensely. The quality of coloration by acid fuchsin varies according to root tissue and fungal species, helping to distinguish different endophytes. Its effectiveness may be greatly enhanced under the fluorescence microscope using green light (Merryweather and Fitter, 1991). Most stains are extremely hazardous (e.g. acid fuchsin, trypan blue, chlorazol black E) and should be used only when safer alternatives, such as aniline blue or methyl blue, are impractical.

Roots may be treated in test tubes, their contents strained through a small sieve between treatments. A semi-automatic method has been devised in which root samples are placed in plastic tubes which have had small holes drilled into their bases. The tubes are set in a stainless-steel rack which is moved from solution to solution carrying a full set of tubes with it. **Do not use aluminium racks which dissolve violently in KOH solutions.** The treatment solutions are contained in stainless-steel reservoirs which contain just enough fluid to cover the roots. In hot staining, these are placed in the water bath (Fig. 12.1).

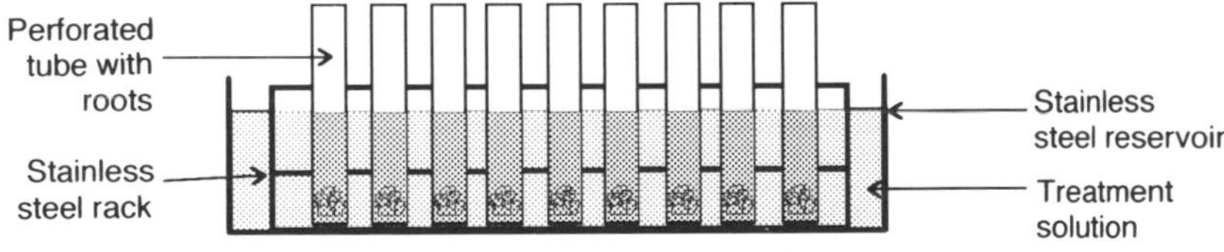

Fig. 12.1. Semi-automatic method for staining arbuscular mycorrhizas.

Hot staining

After Kormanik and McGraw (1982), but see Brundrett *et al.* (1994) for alternatives.

This method is rapid and, apart from destaining, is completed within a few hours.

- Wash root sample and transfer to labelled tube.

- Clear in 10% potassium hydroxide solution in a water bath at 90°C in a fume cupboard. Clearing removes cytoplasm from host root cells, leaving root structure and fungal elements intact.
- The length of time required varies greatly between species. White, fleshy roots may take less than 10 min, others substantially longer. If roots are left in hot KOH for too long, the cortex will disintegrate. Some grass species may lose their cortex before clearing adequately at 90°C. Brief treatment in an autoclave (120°C) may help. Darkly pigmented roots may need to be bleached with hydrogen peroxide at room temperature for between 30 min and several hours. Both alkaline and acidic H_2O_2 have been found to work for different roots.
- Drain and wash thoroughly. Acidify cleared roots by immersion in 0.1 M hydrochloric acid for at least 1 h.
- Stain for 10-30 min in a water bath at 90°C in a fume cupboard in 1% acid fuchsin stain (dissolve 0.01% (w/v) acid fuchsin in the destaining solution of 14:1:1 lactic acid:glycerol:water). Some roots stain well if the rack is lifted and replaced periodically to refresh stain in the tubes.
- Drain and wash thoroughly.
- Destain overnight or longer to remove coloration from empty root cells, etc.
- Mount roots in fresh destaining solution on a microscope slide beneath a 50 × 22 mm No. 1 cover slip. Roots should ideally be mounted as parallel lengths, avoiding tangles, which make quantification very difficult.

Cold staining

(After Koske and Gemma, 1989; Grace and Stribley, 1991; Walker and Vestberg, 1994.) This method takes longer to do, but requires much less attention than hot staining. Large containers may be used to hold solutions allowing many root samples to be processed at the same time. Aniline blue is recommended because, to date, it has not been proven hazardous, unlike most other compounds used for staining fungi (e.g. acid fuchsin, chlorazol black E).

- Wash root sample and transfer to labelled tube.
- Clear in 20% (w/v) potassium hydroxide solution for 1-3 days (It is necessary to experiment to discover the optimal clearing times.)
- Drain, wash thoroughly and acidify with 0.1 M hydrochloric acid.
- Stain with 0.05% (w/v) aniline blue solution for 1-3 days (0.25 g aniline blue dissolved in a destaining solution of 25 ml water and 475 ml lactic acid).
- Destain overnight or longer.
- Mount roots in fresh destaining solution (see above) on a microscope slide.

Quantification of Arbuscular Mycorrhizas

The magnified intersection method (McGonigle *et al.*, 1990)

- Place the slide on the stage of a compound microscope equipped with a crossed-hair eyepiece graticule. *(Assessment of mycorrhizal colonization using a dissecting microscope is likely to be inaccurate due to the difficulty of differentiating mycorrhizal fungi and others at low magnifications. It is essential to confirm that fungi are arbuscular to obtain meaningful results.*)
- Scan the slide methodically, one axis of the graticule aligned with the long axis of each root encountered (Fig. 12.2).
- For a simple count of percentage of root length colonized by mycorrhizal endophyte (% RLC), score presence or absence of arbuscular endophyte touched by the graticule axis which crosses the root each time a root is encountered.
- One may also score number of hyphae encountered (up to the point where hyphal density prevents accurate counting, when a maximum score must be recorded), number of entry points, arbuscule numbers and vesicles encountered by the graticule axis.
- It is possible to achieve a reasonable estimate of % RLC from 25 intersections per slide, but accuracy is achieved only if 100 or more counts are recorded.
- Counts are recorded as percentages:

$$\% \text{ RLC} = \frac{100 \times \text{Number of intersections with arbuscular hyphae}}{\text{Total number of intersections counted}}$$

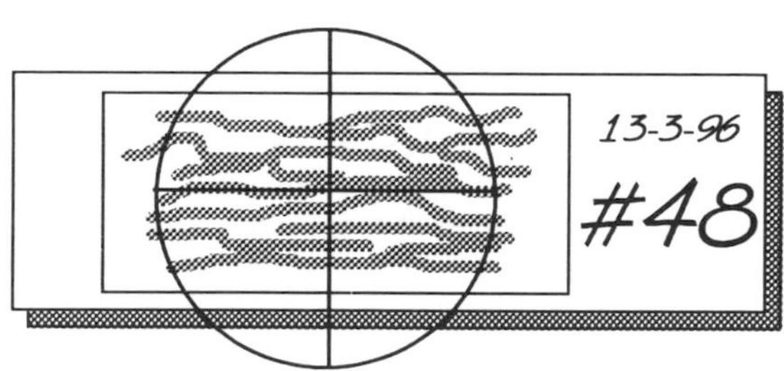

Fig. 12.2 The magnified intersection method.

Limitations of Existing Techniques

The central problem in AM research is that the fungi are obligate symbionts. This means that individual isolates must be maintained in pot culture. It is essential that all workers retain specimens of taxa they have worked on, ideally in single-spore culture. These can then be characterized by molecular techniques at one of the various laboratories that have the appropriate skills. Only by this means can an accurate and usable taxonomy evolve, which will permit the biology of the fungi in symbiosis to be elucidated.

It is often difficult to elucidate the identity and dynamics of the fungi in AM symbioses on morphological grounds. The taxonomy of arbuscular mycorrhizal fungi is based on the morphology of spores obtained from pot-grown symbioses; identification of field-collected spores is difficult. This limits ecological studies, since sporulation depends on environmental conditions and both host and fungal genotype. Ecological studies therefore require the assumption that the number of spore types extracted from a soil sample reflect the colonization of fungi in host roots (Morton *et al.*, 1995), an assumption that is now being questioned by the use of molecular methods of identification (Simon *et al.*, 1992a) and can give a very unreliable indication of root colonization in the field (Clapp *et al.*, 1995).

Suggestions to Overcome Technical Limitations

Ectomycorrhizal fungi are easier to identify because of their large and diverse sporophores. However, in ecological studies it is often necessary to identify extramatrical hyphae which is only possible using molecular techniques. Many techniques being used predominantly for ectomycorrhizal research may be equally applicable to endomycorrhizas. Without exception, all these techniques utilize the ability of the polymerase chain reaction (PCR) to amplify DNA from extremely dilute or crude samples having a preponderance of non-target DNA or of non-DNA substances.

DNA Extraction

A prerequisite is to obtain DNA from the samples. DNA can be extracted from both fungal spores and mycorrhizal roots and it is usually possible to obtain amplifiable DNA from single spores (Wyss and Bonfante, 1993). We use a 'Chelex' resin (BioRad) DNA extraction method (Walsh *et al.*, 1991) regularly.

Various buffers are used to extract DNA from mycorrhizal roots. They usually contain a combination of CTAB (hexadecyltrimethylammonium bromide), SDS (sodium dodecyl sulphate), β-mercaptoethanol or Proteinase K and involve an incubation at 50-65°C for 30-60 min (e.g. Gardes *et al.*, 1991; Henrion *et al.*, 1992; Lanfranco *et al.*, 1995). The numerous variations appear to be based on personal preference and experience will decide which one is most effective in any particular laboratory. Homogenization of material within the extraction buffer will suffice on most occasions. Homogenization with liquid nitrogen prior to extraction is only really necessary when very small quantities of root material are being used or quantification of DNA is being done (as other methods of homogenization are less standardized). The root homogenate is usually incubated with phenol to remove protein and then nucleic acids are retrieved by alcohol precipitation.

Genus-Specific Primers and PCR

Primers specific to the order *Glomales* (VANS1) and to the genera *Gigaspora/Scutellospora* (VAGIGA), *Acaulospora* (VAACAU) and *Glomus* (VAGLO) have been used in conjunction with PCR to successfully amplify fungal DNA from the heavily colonized roots of laboratory-grown plants (Simon *et al.*, 1992a,b; Simon, 1994). In our experience, the genus-specific 18S ribosomal RNA (rRNA) gene primers, although successful on AM-colonized onion roots from glasshouse cultures, failed to produce amplified products directly from AM roots collected in the field. We have found, however, that it is relatively simple to amplify products from field-collected roots using the universal rRNA gene primers SS38 (Bousquet *et al.*, 1990) and NS21 (Simon *et al.*, 1992a), which indicates that the failure of the genus-specific primers in field samples is not due to inhibition of the amplification itself. Even so, subsequent amplification of the universal products with the genus-specific primers still fails to yield products. This may be due to inhibition of the specific amplification due to a superabundance of host plant DNA in root extracts, and may not occur in all plant-fungus combinations. We have developed a method that uses subtractive hybridization (Clapp *et al.*, 1993) to enrich for low-abundance, fungus-derived sequences after an initial amplification using the universal primers for the 18S rRNA gene. This technique which we have called Selective Enrichment of Amplified DNA (SEAD), assumes that a subset of the products derived from such an amplification is of fungal origin but inaccessible to the genus-specific primers. We have used it to determine the mycorrhizal taxa in the roots of bluebell *Hyacinthoides non-scripta* plants collected from field populations (Clapp *et al.*, 1995).

Random Amplified Polymorphic DNA (RAPD) Analysis

RAPD analysis is a variant of the standard PCR technique. It uses single, short, arbitrary primers which anneal frequently throughout the genome. Where the primers anneal in inverted orientation on opposite stands within approximately 4 kb of each other, amplifiable products can be produced. Usually several products of varying size are formed, which can be visualized by agarose gel electrophoresis and ethidium bromide staining. DNA differences that disrupt or create primer annealing sites will determine the pattern of products visualized. In this manner DNA polymorphism can be detected without specific nucleotide sequence information. Wyss and Bonfante (1993) used RAPD analysis to show polymorphism between and within AM fungal species. Lanfranco *et al.* (1995) used RAPDs to isolate species-specific primers for use in standard PCR. In this case, a RAPD band specific for *Glomus mosseae* was identified and sequenced, then used to design primers which amplified the *G. mosseae* RAPD fragment alone. Since this method is not based on sequence comparisons, the specificity of the amplified product must be determined empirically.

PCR-Restriction Fragment Length Polymorphism (PCR-RFLP) of the Internal Transcribed Spacer (ITS) Regions

The ITS region is being used extensively for the identification of extramatrical mycelia of ectomycorrhizal fungi. It lies between the genes for 18S and 26S ribosomal RNA and is divided into ITS1 and ITS2 by the 5.6S gene. PCR primers are used to amplify across the region and the amplification product digested using restriction enzymes with four base recognition sites. This generates a genus- or species-specific pattern on an agarose gel stained with ethidium bromide. The primers ITS1 and ITS4 are situated in conserved regions at the 3' end of the 18S and 5' end of the 26S genes and as such are universal in their specificity. This leads to difficulties when the template is mixed, but these have been largely overcome by the development of primers with enhanced specificity for basidiomycetes - ITS1-F and ITS4-B (Gardes and Bruns, 1993).

The ITS is being explored with a view to assessing AM diversity by PCR-RFLP. Currently however, the intergenic region (IGR) comprising the intergenic spacer (IGS) and external transcribed spacer (ETS) is thought to have even greater potential for obtaining identifiable sequence variation. The techniques required for PCR-RFLP are described in detail in Gardes and Bruns (1996).

Single-Stranded Conformational Polymorphism (SSCP)

This technique has not been greatly exploited in the study of mycorrhizas. It is most useful for rapidly screening amplified sequences from large numbers of isolates for variation (Simon *et al.*, 1993) and is particularly suitable for assessing intraspecific variation. It has been reported to be capable of detecting single base substitutions in moderately sized PCR products (up to 450 base pairs). A full description of this technique as applied to AM fungi is given by Simon (1994).

Quantification of Arbuscular Mycorrhizal Fungi Within the Host

Traditional methods of quantifying AM fungi within roots are based on either chemical analysis or microscopic counting. The first method involves the analysis of total sterol or chitin content from all fungal sources within roots, and is not specific to AM fungi. A PCR-based technique which allows the quantification of AM fungi within roots is fast and can be specific to order, genus or species. Simon *et al.* (1992b) have developed such a quantitative method using competitive PCR with the order-specific primer pair VANS1 and

NS21. This method is based on a comparison made between an artificially constructed internal standard (IS) and the amplified AM product from an infected root. The internal standard is produced by removing a central section from a normal product and re-ligating the termini. The resultant molecule has identical priming sites to that of a molecule amplified from an infected root but is slightly smaller. The internal standard is cloned into a plasmid vector and added in known quantity to a root extraction. By determining the relative intensity of staining of the native and IS products by agarose gel electrophoresis, an estimate of AM colonization of a root can be made. When the ribosomal copy numbers of AM fungi are known, this method will give a very precise way of analysing AM colonization of roots.

Integrating Techniques into a General Scheme of Analysis

For native analysis, visual methods of identification will continue to be required, but these require to be calibrated against cultures that have been subjected to molecular analyses. It is possible to distinguish some different fungal taxa in roots by morphological criteria (Abbott, 1982), but at present only after the fungi have been brought into culture. Most ecological work is likely to remain dependent on spores isolated from soil, even though these are known not to reflect root populations accurately (Clapp *et al.*, 1995). Where possible, voucher specimens of spore types should be preserved in a form (e.g. flash-frozen in liquid nitrogen and stored at -80°C) that makes subsequent molecular characterization possible.

Bibliography

General and Specific Texts

Abbott, L K. (1982) Comparative anatomy of vesicular-arbuscular mycorrhizas formed on subterranean clover. *Australian Journal of Botany* 30, 485-499.

Brundrett, M., Melville, L. and Peterson, L. (1994) *Practical Methods in Mycorrhiza Research*. Mycologue Publications, Guelph.

Clapp, J.P., McKee, R.A., Allen-Williams L., Hopley, J.G. and Slater RJ. (1993). Genomic subtractive hybridisation to isolate specific DNA sequences in insects. *Insect Molecular Biology* 1, 133-138.

Clapp, J.P., Young J.P.W., Merryweather, J.W. and Fitter, A.H. (1995) Diversity of fungal symbionts in arbuscular mycorrhizas from a natural community. *New Phytologist* 130, 259-265.

Grace, C. and Stribley, D.P. (1991) A safer procedure for routine staining of vesicular-arbuscular mycorrhizal fungi. *Mycological Research* 95, 1160-1162.

Kormanik, P.P. and McGraw, A.-C. (1982) Quantification of vesicular-arbuscular mycorrhizae in plant roots. In: Schenck, N.C. (ed.) *Methods and Principles of Mycorrhizal Research.* American Phytopathological Society, St Paul.

Koske, R.E. and Gemma, J.N. (1989) A modified procedure for staining roots to detect VA mycorrhizas. *Mycological Research* 92, 486-505.

McGonigle, T.P, Miller M.H, Evans, D.G, Fairchild, D.G and Swann, J.A. (1990) A new method which gives an objective measure of colonisation of roots by vesicular-arbuscular mycorrhizal fungi. *New Phytologist* 115, 495-501.

Merryweather, J.W. and Fitter, A.H. (1991) A modified method for elucidating the structure of the fungal partner in a vesicular-arbuscular mycorrhiza. *Mycological Research* 95, 1435-1437.

Morton, J B. (1995) Impact of host species on the culture of arbuscular fungi. *INVAM Newsletter* 5, 4-6.

Morton, J.B., Bentivenga, S.P. and Bever J.D. (1995) Discovery, measurement, and interpretation of diversity in symbiotic endomycorrhizal fungi (*Glomales, Zygomycetes*). *Canadian Journal of Botany* 73 (Suppl. 1), Section A-D, S25-S32.

Trappe, J.M. (1987) Phylogenetic and ecologic aspects of mycotrophy in the angiosperms from an evolutionary standpoint. In: Safir, G.R. (ed.) *Ecophysiology of VA Mycorrhizal Plants.* CRC Press, Boca Raton, pp. 5-25.

Walker, C. and Vestberg, M. (1994) A simple and inexpensive method for producing and maintaining closed pot cultures of arbuscular mycorrhizal fungi. *Agricultural Science in Finland* 3, 233-240.

Molecular Biology of Mycorrhizas

Bousquet, J., Simon, L. and Lalonde, M. (1990) DNA amplification from vegetative and sexual tissues of trees using polymerase chain reaction. *Canadian Journal Forestry Research* 20, 254-257.

Bruns, T.D. and Gardes, M. (1993) Molecular tools for the identification of ectomycorrhizal fungi-taxon-specific oligonucleotide probes for suilloid fungi. *Molecular Ecology* 2, 233-242.

Erland, S., Henrion, B., Martin, F., Glover, L.A. and Alexander, I.J. (1994) Identification of the ectomycorrhizal basidiomycete *Tylospora fibrillosa* Donk by RFLP analysis of the PCR-amplified ITS and IGS regions of ribosomal DNA. *New Phytologist* 126, 525-532.

Gardes, M. and Bruns, T.D. (1993) ITS primers with enhanced specificity for basidiomycetes - application to the identification of mycorrhizae and rusts. *Molecular Ecology* 2, 113-118.

Gardes, M., and Bruns, T.D. (1996) ITS-RFLP matching for identification of fungi. In: Clapp, P.J. (ed) *Methods in Molecular Biology*. Humana Press, New Jersey, pp. 177-186.

Gardes, M., White, T.J., Fortin, J.A., Bruns, T.D. and Taylor, J.W. (1990) Identification of indigenous and introduced symbiotic fungi in ectomycorrhizae by amplification of nuclear and mitochondrial ribosomal DNA. *Canadian Journal of Botany* 69, 180-190.

Gardes, M., Mueller, G.M., Fortin, J.A. and Kropp, B.R. (1991) Mitochondrial DNA polymorphisms in *Laccaria bicolor, L. laccata, L. proxima* and *L. amethystina. Mycological Research* 95, 206-216.

Henrion, B., Le Tacon, F. and Martin, F. (1992) Rapid identification of genetic variation of ectomycorrhizal fungi by amplification of ribosomal RNA genes. *New Phytologist* 122, 289-298.

Lanfranco. L., Wyss, P., Marzachi, C. and Bonfante, P. (1995) Generation of RAPD-PCR primers for the identification of isolates of *Glomus mosseae*, an arbuscular mycorrhizal fungus. *Molecular Ecology* 4, 61-68.

Norris, J.R., Read, D.J. and Varma, A.K. (eds) (1992) *Techniques for the Study of Mycorrhiza.* Methods in Microbiology, Vol. 24. Academic Press, New York.

Simon L. (1994) Specific PCR primers for the identification of endomycorrhizal fungi. In: Clapp, J.P. (ed.) *Methods in Molecular Biology*. Humana Press, New Jersey.

Simon, L., Lalonde, M. and Bruns, T.D. (1992a) Specific amplification of 18S fungal ribosomal genes from vesicular-arbuscular endomycorrhizal fungi colonising roots. *Applied and Environmental Microbiology* 58, 291-295.

Simon L, Lévesque, R.C. and Lalonde, M. (1992b) Rapid quantitation by PCR of endomycorrhizal fungi colonising roots. *PCR Methods and Applications* 2, 76-80.

Simon, L., Lévesque, R.C. and Lalonde, M. (1993) Identification of endomycorrhizal fungi colonising roots by fluorescent single-strand conformational polymorphism-polymerase chain reaction. *Applied and Environmental Microbiology* 59, 4211-4215.

Walsh, P.S., Metzger, D.A. and Higuchi, R. (1991) Chelex 100 as a medium for simple extraction of DNA for PCR-based typing from forensic material. *Biotechniques* 10, 506-513,

Wyss, P. and Bonfante, P. (1993) Amplification of genomic DNA of arbuscular-mycorrhizal (AM) fungi by PCR using short arbitrary primers. *Mycological Research* 95, 1351-1357.

Useful Addresses

BEG, c/o Dr John C. Dodd, International Institute of Biotechnology, Research School of Biosciences, PO Box 228, Canterbury, Kent CT2 7YW, UK.

INVAM, Division of Plant and Soil Sciences, 401 Brooks Hall, Box 6057, West Virginia University, Morgantown, WV 26506-6057, USA.

Micro- and Macro-Arthropods 13

John E. Bater

IACR-Rothamsted, Harpenden, Hertfordshire AL5 2JQ, UK.

Soil Arthropods and Ecosystem Processes

Scientists have long held the activity of animals to be a powerful factor in soil formation. Since the early work of soil scientists there had been a rather long period of accumulation of data on animals inhabiting the soil and on their population density before soil zoology could develop. Today, at our disposal, are data enough concerning the hidden world of the soil invertebrates, the studies of which, and practical conclusions drawn from their results, unify the interests of soil zoologists from around the world.

In all terrestrial ecosystems the overwhelming majority of invertebrate species and the bulk of individuals are soil dwellers or they are closely connected with the soil at some period of their development cycle. The classical estimations of Buckle (1923) of invertebrate species associated with the soil put the figure as high as 95-98%. It is also known that the general number of insect species is much over one million. The species-rich group of mites is also represented chiefly by soil dwellers. Practically all myriapods are represented by soil inhabitants. Work carried out in the former USSR (Ghilarov and Chernov, 1975) also showed that the majority of invertebrate species in all undisturbed ecosystems were soil dwellers.

As to the numbers of individuals of various groups of invertebrates in the soil the results of counts are very impressive. Even such a simple method as hand soil sifting and sorting samples shows that in various types of soil, under different vegetation, the amount of larger invertebrates, such as earthworms, centipedes, insect larvae, etc., varies between tens and hundreds per m^2. Using even the incomplete but simple methods of various modifications of the

'Tüllgren' funnel one can find that in the soil there are thousands or even hundreds of thousands of individuals belonging to so-called microarthropods.

The soil represents a very complicated and many-sided environment. It is the stratum of biogeocenoses where transformation, decomposition, mineralization and humification of organic matter takes place, the very focus of matter turnover in terrestrial ecosystems. Dying-off roots, fallen leaves and other litter, dead bodies and faeces of all terrestrial animals, all dead remnants of terrestrial organisms become the substrate attacked by all groups of microorganisms. Therefore, all stages of decomposition of plant and animal remnants are to be met in the soil, and consumers specialized on decaying matter of various origins and at different stages of decomposition can simultaneously find their food in the soil. Plant eaters can attack roots and other underground parts of various plants penetrating the soil and all groups of consumers and reducers represent links in food chains of predators and parasites and their carcasses become food of necrophagous and sapro-necrophagous animals.

Therefore, in the soil representatives of various ecological groups can be found, from the physiologically aquatic to the physiologically strictly terrestrial ones with the whole spectrum of transition, of all types of feeding habits, and a wide range of body sizes. Thus, representatives of many taxa and of the majority of phyla can be found in the soil with a great multitude of species diversity.

The complex composition of soils with varying pores, passageways and immense surface area, allows soil invertebrates access to the cavities between soil aggregates. They find in the soil appropriate environment, food and shelter, enabling protection from desiccation and from predators. The possibility to migrate up and down within the soil using existing cracks and cavities allows them to find an appropriate stratum with the necessary preferred hydrothermal regime. Therefore, not only eurybiotic but also stenobiotic small invertebrates can co-exist in the soil.

For macroinvertebrates, the whole soil represents their habitat, but the mode of use of this environment may be very different in various taxa and ecological groups. Even for larger invertebrates, the soil can represent a system of passages used in their migration (geophilids), or a loose substrate in which it is possible to move (as in the case of earthworms or many dipterous larvae) or a rather solid substrate in which it is only possible to dig or bore ways in search of food and shelter, as many insect larvae do (Ghilarov, 1949).

In terrestrial ecosystems there is no other stratum comparable to the soil with possibilities to correspond to so many quite different ecological demands of various groups of animals, where the permanent renewal of primary and secondary production supplies takes place. It should be stressed that the soil represents quite different environments for various groups of animals: small water bodies for protozoa, rotifers and small nematodes, a system of caves, passageways and cracks for microinvertebrates, and for macroinvertebrates it represents an almost solid substrate. So, in the soil there are various animal worlds co-existing in quite different ecological conditions. The complexity of both the soil matrix and the invertebrates that live, reproduce and die in and on this substrate, together with the very high numbers and diversity of species existing per unit of soil, presents an ideal opportunity to try and unravel the complex relationships that exist between and among soils and soil invertebrates. This information can be used to understand the functioning of terrestrial and marine ecosystems by inventorying and monitoring biodiversity, and the effects of both natural and man-made agents on both the soil and its associated invertebrates.

Sampling/Extraction Techniques

Soils differ greatly in composition, particle size, structure, depth and compaction, and whether they are under trees, grassland or cultivation. The invertebrates that inhabit these soils are extremely diverse in form, differ greatly in size, occur in extremely large numbers, are more or less aggregated in distribution, both vertically and horizontally, and often live deep within soil particles. Since the soil fauna is incorporated closely into the soil structure, the assessment of populations of these organisms is extremely difficult and laborious and necessitates a wide range of specialized techniques if populations of all of the wide range of types of organisms are to be assessed (Edwards, 1991).

A difficult task is made even more complex by the seasonal changes in the invertebrate population and their position in the soil stratum. (Leinaas, 1978; Takeda, 1979; van Straalen and Rijninks, 1982). The life stages of invertebrates will also change seasonally with many changing from a sessile mode to a highly mobile one. Some invertebrates will penetrate very deep into the soil at certain times of the year, making sampling almost impossible. Therefore, to study populations of soil invertebrates thoroughly will necessitate sampling at different times during the year and, at the very least, sampling should take place in spring and autumn when invertebrate activity is at its highest. Sampling for invertebrates in the tropics will be much the same as sampling in temperate regions, although seasons will, of course, vary and extremes such as prolonged dry or monsoon periods should be avoided. Sampling should be carried out at equivalent times to spring and autumn in temperate regions. The following section looks at the various types of sampling techniques available and which are most commonly used, together with the limitations of the technique, and where possible, how these limitations may be overcome.

Field Population Assessment Methods

Field counting methods

The simplest method to assess populations of macroinvertebrates is to remove a measured quantity of soil from the sample area, which has been delimited by either a quadrat or a circle and to gradually crumble the sample over a sheet of plastic or other material and collect the invertebrates as they are released. They can then be stored in a suitable preservative. Because of the great variability of the invertebrates being looked at, care must be taken in choosing the size of quadrat and depth. The limitations of this simplest of methods is that it is only suitable for invertebrates that can be seen with the naked eye; it is also labour-intensive and time-consuming.

These limitations have been reduced in some instances by, for example, crumbling the soil into a container of water, where many smaller invertebrates will float to the surface and can be picked off. Additions of water or dilute chemicals onto quadrats will help bring some beetles to the surface and earthworms will certainly be extracted using chemical irritants such as potassium permanganate (Evans and Guild, 1947) or formaldehyde (Raw, 1959). These additions, however, are affected by the seasons. Electrical stimulation of the soil using a generator has also been used, particularly for earthworms (see James, S.W., this volume Chapter 20), but is dependent on soil moisture and pH, and it is difficult to assess the area being sampled.

Field trapping

The various types of traps used are dependent on accidental encounter by the invertebrates, or the use of attractants.

Pitfall traps

The method of sinking containers into the soil, flush with the soil surface and containing a preservative or left dry, is probably the most commonly used method of catching invertebrates. Many different designs are to be found, small or large, usually plastic, containers, gutter traps (lengths of domestic house guttering sunk into the ground), traps set at different depths in the soil and traps that are motorized allowing temporal trapping to be carried out (Ayre and Trueman, 1974; Blumberg and Crossley, 1988). Addition of guiding fins allows directional trapping and also gutter traps, if laid side by side, will indicate approximately the direction in which the insects were travelling. They can be disguised by covering with soil or litter and most usually have a plastic rain guard cover (Fig. 13.1). Traps buried at different depths in the soil can be used to assess vertical distributions of invertebrates.

The limitations of pitfall traps are primarily in the interpretation of data, since the numbers of animals trapped are related to both overall numbers present and their activity, and so may not sample each population entirely. Methods are available to convert the numbers of invertebrates trapped to populations, usually based on physically delimiting sampling areas with some form of barrier and using mark and recapture techniques to relate numbers of invertebrates trapped to numbers that are present in the whole population (Dobson, 1962). Other problems include preservatives used and their toxicity to vertebrates attempting to drink the fluid. This can be controlled quite well by using chicken-wire cylinders around the traps. Length of time of trapping, evaporation of preservative and disturbance of the trap and contents by inquisitive vertebrates and birds can also pose problems.

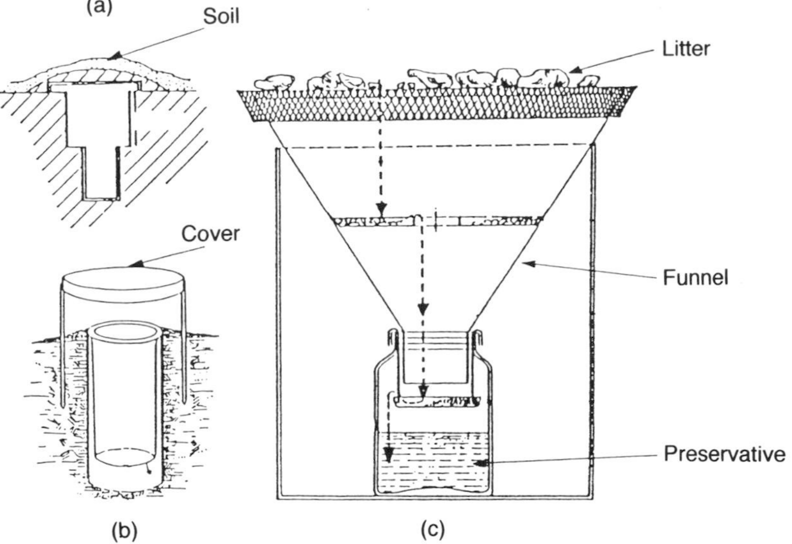

Fig. 13.1. Various kinds of pitfall traps. (a) Soil-covered trap; (b) standard pitfall trap; (c) litter-covered funnel trap.

Baits

Many soil-inhabiting invertebrates are attracted to various baits, placed either in containers at or below the soil surface or buried in soil in mesh bags. Bran or beer baits have been used commonly to attract slugs, tipulid larvae and cutworms (Edwards and Heath, 1964). Leaf tissue, straw or filter paper enclosed in mesh bags has been used to attract a wide range of small soil-inhabiting invertebrates, particularly those belonging to decomposer groups (Blair *et al.*, 1991). Although these techniques are useful to identify those taxa involved in the decomposition process, they are poor for population assessment.

Soil Sampling Methods

Separation of invertebrates from soil samples

There are two main methods of achieving separation of soil invertebrates from the soil in which they live.

- Physical methods of separating the animals from the soil, which involve first wet or dry fragmentation of the soil, followed by flotation onto the surface of dense liquid media, the utilization of differential settling rates of soil particles and invertebrates in flowing water, or techniques based on the greater wetting capacity of the cuticle of arthropods compared with soil particles.
- Dynamic methods which involve exerting environmental or chemical pressures on invertebrates to stimulate them to leave the soil sample under their own volition.

Physical methods are more efficient at recovering all life stages of the invertebrates, including eggs, cysts, pupae and various resting stages.

These methods will also recover recently dead individuals but the methods also produce more damaged specimens, which can be difficult to identify. Most of these physical methods are laborious and time-consuming, limiting the number of samples that can be processed.

Dynamic methods extract invertebrates in very good condition, but can give biased results, since they depend on the activity of the animal and its ability to respond to the stimuli and free itself from the soil. The efficiency will also vary seasonally.

Physical methods

Most physical methods of separating animals from soil depend either on differences in specific gravity between the invertebrates and the soil in which they live, so that they can be separated in a liquid phase, or on the lipophilic and hydrophobic properties of the arthropod cuticle. Once the invertebrates are freed from the soil particles in which they live, they will float to the surface of any solution with a specific gravity of >1.2. Solutions commonly used for such flotations include brine, sugar or magnesium sulphate.

Soil washing and flotation

When soil is dropped into a dense solution instead of water, all the animals that are not trapped in the soil float to the surface and can be picked off or decanted. Unfortunately, most of the organic matter in soil also floats to the surface and efficiency of recovery can be greatly increased by washing the soil through a series of sieves with decreasing mesh size to separate out the stones and unwanted organic debris prior to flotation (Morris, 1922). This efficiency can also be increased in other ways using organic solvents (Ladell, 1936; Salt and Hollick, 1944; Calvert, 1987). In some countries, however, the use of solvents such as xylene and benzene may be restricted or even banned, and so it is important that local health and safety regulations are consulted. If alternatives are required, then it will be necessary to liaise with chemists to find the best alternatives. Other mechanized versions have been developed that agitate, break up and separate out the debris prior to flotation and reduce the processing time of samples (Edwards and Dennis, 1962).

More recently, a heptane flotation method based on similar principles has been developed by Walters *et al.* (1987). Comparisons indicate a high efficiency for this method, which has been further refined by Geurs *et al.* (1991). Other methods, such as elutriation (see Chapters 15, 18), will extract soil arthropods, but were developed primarily to extract nematodes.

Limitations on these physical methods tend to be time and labour in order to process large numbers of samples.

Dynamic methods

These methods are based on taking a soil sample and applying a physical or chemical stimulus to drive the invertebrates from the sample into a collecting fluid (Southwood, 1966; Edwards, 1991).

Dry-funnel methods. First developed by Berlese (1905), the soil or litter sample rested on a wire mesh and the emerging animals fell into a tube beneath. This was subsequently modified by Tüllgren (1918), who used heat from an

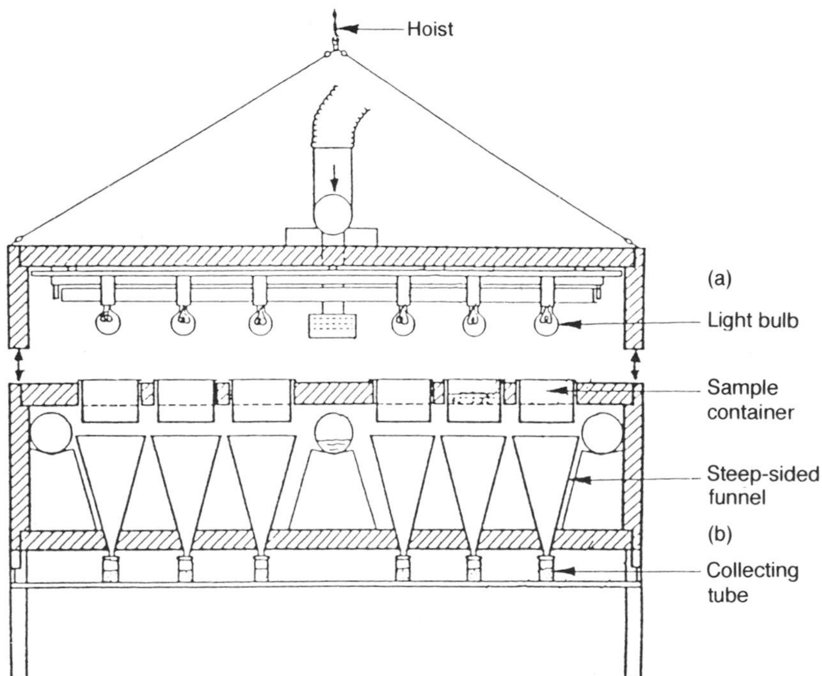

Fig. 13.2. Rothamsted-modified Macfadyen high-gradient funnel. (a) Removable lid lifted by car jack; (b) main body of apparatus.

electric bulb suspended over the sample to drive the ivertebrates from the soil sample.

The Tüllgren funnel is now the most widely used extraction apparatus, many of which have their own specific modifications, e.g. the Rothamsted-modified Macfayden high-gradient funnel (Fig. 13.2). All the successful designs and modifications depend on a carefully controlled high gradient of temperature and humidity between the top and the bottom of the sample (Nef, 1962; Edwards and Fletcher, 1971; Takeda, 1979; Bieri *et al.*, 1986; Crossley and Blair, 1991). This design has also been copied on small portable funnels (Usher and Booth, 1984). The main drawback of these type of extractors has been the amount of debris and soil that falls into the sample vial, causing increased processing time. Efforts to alleviate this have proved successful but, at the same time, have reduced the efficiency of the apparatus.

Wet-funnel methods Many smaller, fragile hydrophilic invertebrates are not extracted efficiently by dry-funnel techniques and, instead, are extracted more efficiently by a number of wet-funnel techniques. Most typical of these is the Baermann funnel (1917; Fig. 13.3a) with a modification by O'Connor (1955; Fig. 13.3b). This method is also very efficient at extracting *Enchytraeidae* from soil and for tardigrades which are not extracted by the dry-funnel method.

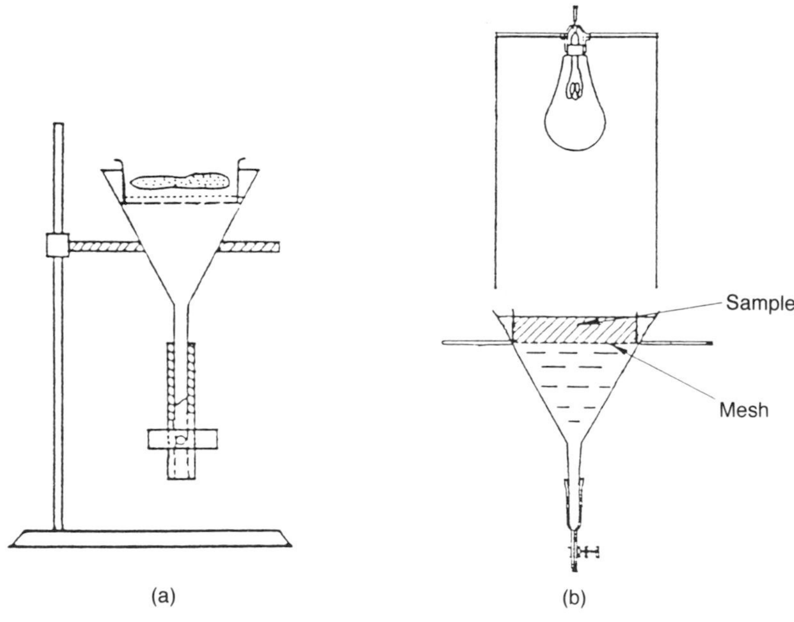

Fig. 13.3. Baermann funnels for nematode, enchytraeid and microarthropod extraction. (a) Original Baermann funnel (Baermann, 1917); (b) modified heated funnel (O'Connor, 1955).

Wet-container methods Baermann's principle has also been put to use in extracting forest litter, using a beaker instead of a funnel and agitating periodically (Mindermann, 1956). This greatly improves the method and makes it much more suitable for microarthropods. Modifications to this method allow improved extraction of *Enchytraeidae* and tipulid (*Diptera*) larvae.

Integration of Techniques into a General Scheme of Analysis

The selection of suitable methods of assessment for soil-inhabiting invertebrate populations is a critical choice in many ecological studies. A comprehensive assessment of the invertebrate populations of even a single site would require a range of different extraction techniques and would be extremely laborious and time-consuming. However, the integration of these methods of assessment of soil invertebrate populations into an overall general assessment should be relatively straightforward.

Careful consideration is necessary of exactly what information is required from a site, i.e. whether a complete invertebrate inventory is needed or specific groups. Once decided, the site will then dictate the number of samples taken; the size of sample and the soil type will dictate to some extent the type of extraction technique required to obtain the best results. No single extraction method or size of sample is best for all groups of animals, and it may be necessary to use several methods and sample sizes. Alternatively, where a single method is used for most of the estimations, its efficiency for each group of animals in that particular site should be investigated before it is used extensively.

To obtain a complete inventory of micro- and macroinvertebrates, a minimum of three techniques would be required. To obtain the bulk of micro-arthropods, the best general method seems to be a Tüllgren funnel of the Macfayden type, with a steep gradient of temperature and moisture through the sample. To obtain eggs, pupae or animals in diapause, a mechanical flotation method is essential; however, the efficiency will depend on the soil type and amount of organic matter present in the sample. Finally, pitfall traps will complete the overall assessment with the capture of most macroinvertebrates. In some instances it may be necessary to include a specialized trap design for specific animals that the aforementioned techniques may not catch or extract readily. For example, in recent years the pear thrips, *Taeniothrips inconsequens* Uzel, has been causing major problems in sugar maple stands in north-eastern USA. A programme was set up to assess the extent of the damage to instigate a complete forest floor management programme. Within the general monitoring, using the standard techniques, specialized traps were designed and included (Bater, 1991). It was hoped that they would not only monitor the pest, but also the general health of the microarthropod population in the soil and would also allow monitoring of any chemical control methods used on the thrips that might also affect other forest floor invertebrates.

It must be emphasized that careful preliminary investigation is needed to assess the numbers of soil arthropods from any site accurately, and that adequate equipment and statistical procedure, very considerable labour and a thorough taxonomic facility are essential.

Bibliography

Adis, J. (1987) Extraction of arthropods from neotropical soils with a modified Kempson apparatus. *Journal of Tropical Ecology* 3, 131-138.

Ayre, G.L. and Trueman, D.K. (1974) A battery operated time-sort pitfall trap. *The Manitoba Entomologist* 8, 37-40.

Baermann, G. (1917) Eine einfache methode zur Auffindung von *Ankylostomum* (Nematoden) Larven in Erdproben. *Mededelingen uit het Geneeskundig Laboratorium te Weltevreden* 41-47.

Bater, J.E. (1991) A trap design for combined insect emergence and soil arthropod extraction from soil. *Agriculure, Ecosystem and Environment* 34, 231-234.

Berlese, A. (1905) Apparecchio per racogliere press ed in gran numero piccoli artropodi. *Redia* 2, 85-89.

Bieri, M., Delucchi, V. and Stadler, M. (1986) Optimization of extraction conditions in the air-conditioned Macfadyen funnel extractor for soil arthropods. *Pedobiologia* 30, 127-135.

Blair, J.M., Crossley, D.A.Jr. and Callahan, L.C. (1991) A litter basket technique for measurement of nutrient dynamics in forest floors. *Agriculture, Ecosystem and Environment* 34, 465-471.

Blumberg, A.Y. and Crossley, D.A.Jr. (1988) Diurnal activity of soil-surface arthropods in agroecosystems: Design for an inexpensive time-sorting pitfall trap. *Agriculture, Ecosystem and Environment* 20, 159-164.

Buckle, P. (1923) On the ecology of soil insects on agricultural land. *Journal of Ecology* 11, 93-102.

Calvert, A.D. (1987) A floatation method using reduced air pressure for the extraction of sciarid fly larvae from organic soil. *Pedobiologia* 30, 39-43.

Crossley, D.A.Jr. and Blair, J.M. (1991) A high efficiency, low technology Tüllgren-type extraction for soil micro-arthropods. *Agriculture, Ecosystem and Environment* 34, 187-192.

Dobson, R.M. (1962) Marking techniques and their application to the study of small terrestrial animals. In: Murphy, P.W. (ed.) *Progress in Soil Zoology*. Butterworths, London, pp. 228-239.

Edwards, C.A. (1991) The assessment of populations of soil-inhabiting invertebrates. *Agriculture, Ecosystem and Environment* 34, 145-176.

Edwards, C.A. and Dennis, E.B. (1962) The sampling and extraction of Symphyla from soil. In: Murphy, P.W. (ed.) *Progress in Soil Zoology*. Butterworths, London, pp. 300-304.

Edwards, C.A. and Fletcher, K.E. (1971) A comparison of extraction methods for terrestrial arthropod populations. In: Phillipson, J. (ed.) *Methods for the Study of Productivity and Energy Flow in Soil Ecosystems.* IBP Handbook No. 18, Blackwell Scientific Publications, Oxford, pp. 150-185.

Evans, A.C. and Guild, J.W.M. (1947) Studies on the relationships between earthworms and soil fertility. I. Biological studies in the field. *Annals of Applied Biology* 34, 307-330.

Geurs, M., Bongers, J. and Brussaard, L. (1991) Improvements of the heptane flotation method for collecting microarthropods from silt loam soil. *Pedobiologia* 34, 213-221.

Ghilarov, M.S. (1949) *Peculiarities of Soil as an Environment and its Significance in the Insect Evolution*. Nauk, Moscow. (In Russian.)

Ghilarov, M.S. and Chernov, Y.I. (1975). *Soil Invertebrates in Communities of Temperate Zone. Resources in the Biosphere. Synthesis of the Soviet Studies for IBPI:* Nauka, Leningrad, pp. 218-240. (In Russian with English summary.)

Kempson, D., Lloyd, M. and Ghelardi, R. (1963) A new extractor for woodland litter. *Pedobiologia* 3, 1-21.

Ladell, W.P.S. (1936) A new apparatus for separating insects and other arthropods from soil. *Annals of Applied Biology* 23, 862-879.

Leinaas, H.P. (1978) Seasonal variation in sampling efficiency of Collembola and Protrura. *Oikos* 31, 307-312.

Mindermann, G. (1956) New techniques for counting and isolating free-living nematodes from small soil samples from oak forest litter. *Nematologica* 1, 216-226.

Morris, H.M. (1922) The insect and other invertebrate fauna of arable land at Rothamsted. *Annals of Applied Biology* 9, 282-305.

Nef, L. (1962) The role of desiccation and temperature in the Tüllgren-funnel method of extraction. In: Murphy, P.W. (ed.) *Progress in Soil Zoology.* Butterworths, London, pp. 169-173.

O'Connor, F.B. (1955) Extraction of enchytraeid worms from a coniferous forest soil. *Nature* 175, 815-816.

Raw, F. (1959) Estimating earthworm populations by using formalin. *Nature* 184 (Suppl. 21), 1662.

Salt, G. and Hollick, F.S.J. (1944) Studies of wireworm populations. I. A census of wireworms in pasture. *Annals of Applied Biology* 31, 52-64.

Southwood, T.R.E. (1966) *Ecological Methods.* Butler and Tanner Ltd, London.

Takeda, H. (1979) On the extraction process and efficiency of Macfadyen's high gradient extractor. *Pedobiologia* 19, 106-112.

Tüllgren, A. (1918) Ein sehr einfacher Ausleseapparat für terricole Tierformen. *Zeitschrift für Angewante Entomologie* 4, 149-150.

Usher, M.B. and Booth, R.G. (1984) A portable extraction for separating micro arthropods from soil. *Pedobiologia* 26, 17-23.

van Straalen, N.M. and Rijninks, D.C. (1982) The efficiency of Tüllgren apparatus with respect to interpreting seasonal changes in age structure of soil arthropod populations. *Pedobiologia* 24, 197-209.

Walters, D.E., Kiethley, J. and Moore, J.C. (1987) A heptane flotation method for recovering microarthropods from semi-arid soils with comparison to the Menchant-Crossley high-gradient extraction method and estimates of arthropod biomass. *Pedobiologia* 30, 221-232.

Mites 14

DON A. GRIFFITHS

38 Alexandra Road, Ash, Surrey GU12 6AH, UK.

Mites in Terrestrial and Aquatic Ecosystems

In general, the acari or much of their structure, at least, will be too small to be seen easily with the naked eye. Therefore, with few notable exceptions, expansion in the study of mites closely paralleled developments in microscopy, namely, improvements in the limits of resolution and in specimen preparation techniques.

Once it became possible to describe and classify mites, their presence in every conceivable ecological niche meant that techniques to aid their collection quickly followed. These early systems were, as to be expected, simple and whilst soil ecologists have subsequently spent many years of research attempting to refine such methods simple systems are still the most effective, and it is these which are emphasized here.

Both in species diversity and in numbers of individuals, the acari fall within the top four or five biotypes most frequently found in the top one metre band of the Earth's crust.

The ecosystems into which the soils and sediments of this band may be arbitrarily divided for collecting purposes are as follows:

- surface humus layer,
- sub-surface soil types,
- rock cracks and crevices,
- marine and freshwater sediments,
- special habitats - algae, coral, etc.

Sample Collection and Mite Extraction

Very early in the planning stage of a research or survey study it is important to decide upon sample numbers and, what is often overlooked, sample size. In making these decisions the parameters which need to be considered are as follows:

- Transportation available for hauling samples out of the sampling area, especially important when collecting in remote territories.
- Do you want to obtain living or museum material?
- What is the likely transportation or shelf-life of the samples?
- Unit of storage - liquid, cold store or deep-freeze?
- Is the available manpower sufficient to process the sample load within the time-scale of the project?

To facilitate such decision making, whenever it is possible, within the framework of the project, some system of concentrating the samples should be considered. It will usually involve some form of macro-sieving, using a battery of sieves and more than one mesh size, and physically shaking or water-washing the material. The sieve fraction in which the maximum faunal variety is expected to remain will form the main sample, but a percentage of each of the other fractions must be retained for extraction.

The majority of soil arthropods, particularly mites, are extremely sensitive to humidity gradients. In general, below an equilibrium relative humidity (ERH) of 60%, mites will die. Many will also succumb at very high humidity, although under these conditions an oxygen deficiency, due to fungal metabolism, is usually responsible.

Humus and subsoil samples of moderate humidity are best placed into a plastic bag, the fabric of which can permit some gaseous exchange. If these samples can be kept cool (5-10°C), they can be held for several weeks before processing. However, processing as soon as practically possible after collection is recommended. If not, samples should be examined at 2-3-day intervals for excessive fungal growth or dryness.

Samples such as those collected by pooter may contain only a little vegetative matter, and, because they have been sucked up in an air flow, will generally be too dry. A piece of damp paper, relative in size to the size of the sample container, should be added to the container.

In general, marine samples are extracted through filter systems at the time of collection (described later). The final subsample into which the fauna is collected is then treated with a preservative fluid, and preserved specimens are later extracted in water in the laboratory.

Samples from Rock Cracks and Crevices

Inhabitants of this niche can best be collected individually by means of an insect-collecting aspirator. Such systems range from hand-held pooters operated by the human lungs or small motorized units (battery-driven) up to large back-pack machines. Alternatively, larger specimens can be brushed directly into a tube.

In the majority of cases, the amount of material will be too small to extract, and mites can be directly picked out, or the sample strained through a series of small laboratory sieves (in air, water or alcohol solutions) before picking out specimens.

Samples from the Humus Layer

The method of collection will vary with the needs of the project. Qualitative samples can be scooped up in suitable batch sizes and held in plastic or paper bags. The latter are preferred where condensation problems will occur. More precise sampling can be done using soil core-samplers, usually a strong metal tube of varying lengths.

The sorting and picking out of specimens from such samples as leaf litter, organic and alluvial soils is physically impossible. Moreover, much of the fauna will remain hidden within discrete microniches and may never be discovered.

The traditional method of extraction for these samples is attributed to Berlese (1905), and is now universally known as the Berlese funnel. It relies upon the setting up of temperature and humidity gradients within the sample, produced by a source of heat placed above it. As conditions become unfavourable for the inhabitants, they are forced out. How these gradients develop and their effect upon the sample's fauna will be dealt with after the apparatus has been described.

Berlese used a metal funnel, in which the sample was suspended upon a sieve. The funnel was surrounded by a hot water jacket. As the sample dries out the fauna evacuate to escape desiccation, mostly moving downwards into the neck of the funnel, to which is attached a collecting tube.

Tüllgren (1918) used the then relatively new invention of the electric light as his source of heat. The light and heat, all coming from above, ensured the downward movement of the fauna, and was much more successful. Because of the work of these two pioneers, such an apparatus, including many subsequent innovations, is still known as a Berlese-Tüllgren funnel.

In its simplest form, one can be made from a steep-sided funnel of any material providing the internal surface is reasonably smooth. A sieve, usually about 2-3 mm mesh, is placed in the funnel, about two-thirds of the way down. A collecting tube is fixed to the bottom of the funnel spout, and an electric light, preferably capable of a variable light setting, is placed above the funnel (Fig. 14.1). Such a funnel is entirely adequate for qualitative work.

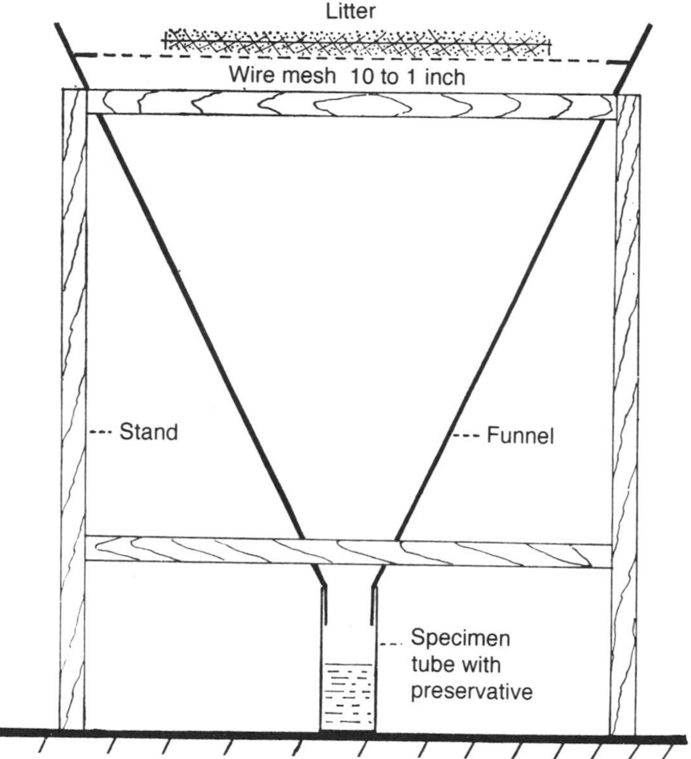

Fig. 14.1. A Berlese-Tüllgren funnel (from Evans *et al.*, 1961).

I have successfully used a Berlese-Tüllgren funnel made from a plastic funnel (top diam of 200 mm) into which is fitted an ordinary kitchen sieve or strainer, with an angle-poise desk lamp as a heat source. Such materials can be bought quite cheaply and, if necessary, made up into a battery of six or more. A similar 'home-made' example (Haarlov, 1947) uses a wooden case, which by closing the lid in steps gives a slow warm-up (Fig. 14.2; see also Chapter 13, Fig. 13.3b).

Whenever using this type of funnel, it is preferable to lower the light source by degrees (in steps) so that animals near the surface of the sample are given time to escape downwards. Where the sample is placed onto a simple grid inserted into the body of the funnel, make sure there is a gap all around the sample between the material and the funnel side. This permits an upward air movement and so prevents condensation on the funnel sides. Likewise, the apparatus needs to be sited in a room with a reasonable ambient temperature.

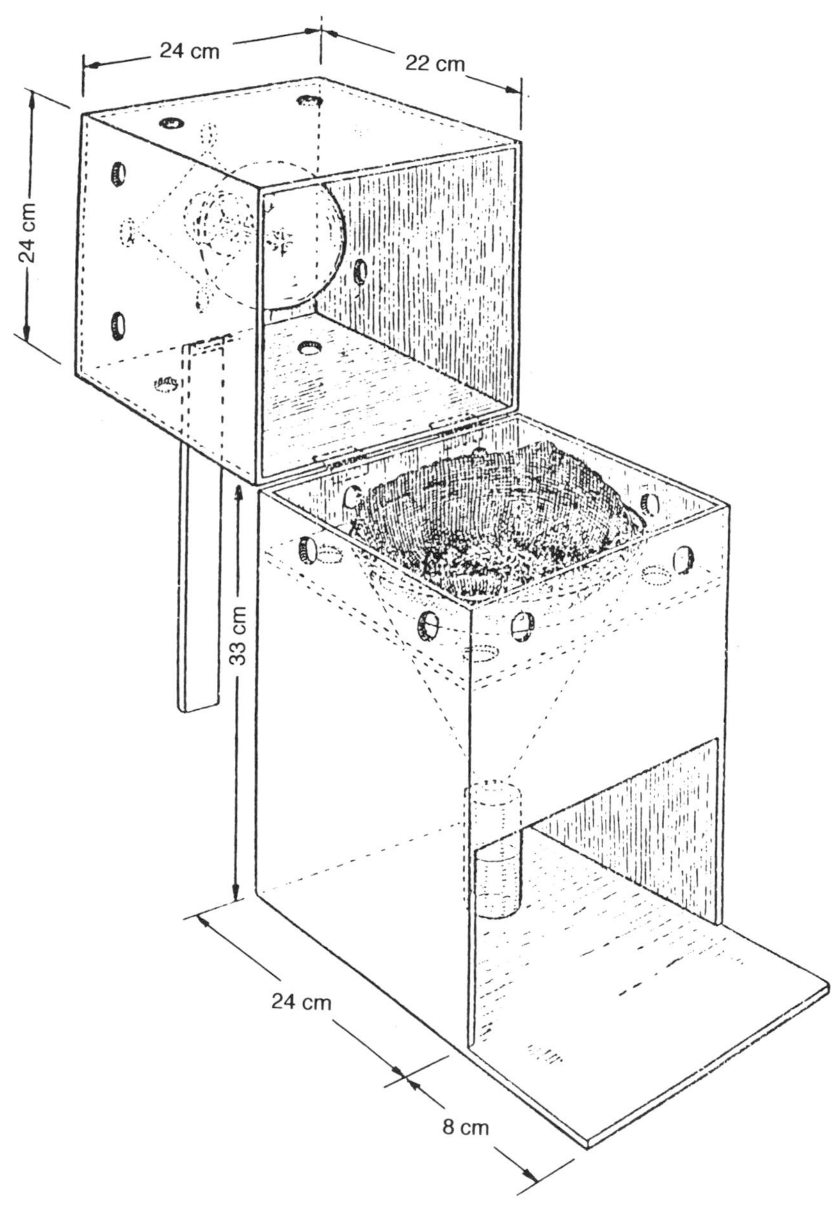

Fig. 14.2 A modified Berlese-Tüllgren funnel (after Haarlov, 1947).

Ideally, a plastic basin of a size which will fit nicely into the body of the funnel should be selected. The bottom is cut off and a gauze of suitable grid size is permanently attached. It can be suspended inside the funnel using a simple plastic collar into which a series of holes have been drilled to permit air movement up between the sample container and the funnel sides. This separate sample container makes it much easier to service the apparatus.

Principles of operation

The temperature gradient in the funnel must be allowed to build up slowly at first to permit those animals near the surface to escape. This initial warm-up period should last overnight and sometimes, particularly in damp material, needs to be longer, 48-72 h. During this period very few specimens may be seen in the collecting tube. The lamp can now be brought down to give a much steeper temperature rise, when species which are temperature-sensitive will descend rapidly, despite the fact that the moisture content in the lower region of the sample is still high (Fig. 14.3).

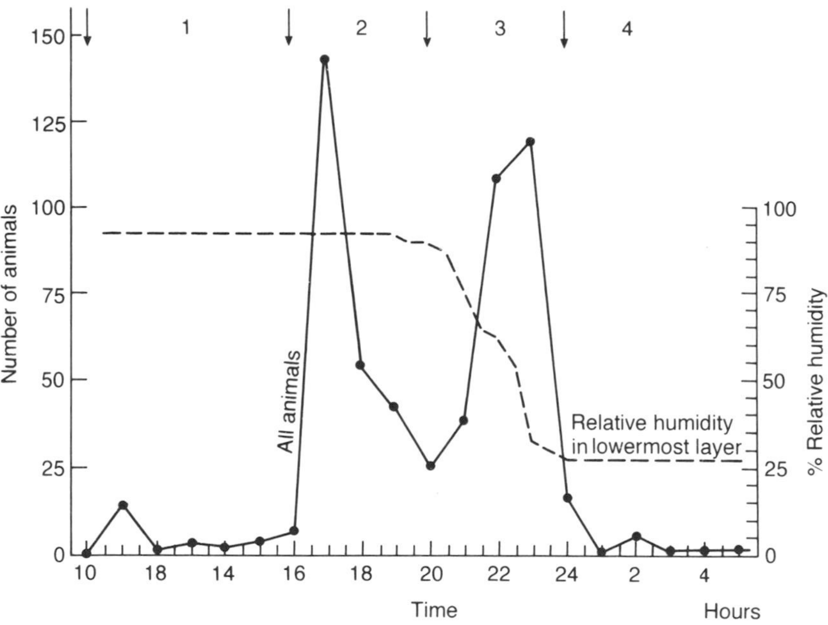

Fig. 14.3 The exodus of mites and collembolids from an extraction apparatus with time (from Haarlov, 1947).

A third wave of escapees will occur when the humidity gradient in the lower half of the funnel becomes unfavourable. The temperature is by now quite high.

There is often a final exodus just before both temperature and desiccation is extreme, so that extraction should not be concluded too soon. In the laboratory, the best way to control the temperature gradient is to fit the light source with a variable rheostat.

Subsoil Samples

The above Berlese-Tüllgren apparatus may be used for all types of loose material, as well as more compacted soil if it is first broken up. However, Macfadyen (1953), closely followed by Murphy (1955), developed high-gradient funnels for quantitative sampling of undisturbed soil cores. The cores are extracted by means of a steel tube, to which a metal gauze lid is fitted over the top of the sample before the core is removed from the ground. This tube can be placed in a second outer container for transportation. To extract the sample the core, still in its tube, is placed in a very steep-sided funnel with the grid downwards, so that mites use the inverted natural channels in the soil as escape routes (see Chapter 13, Fig. 13.2).

Flotation extraction

High mineral content soils are often extracted using flotation-extraction systems. A series of apparatus have been developed which use a mixture of water and a wetting agent, or a water-benzol/ether interface (Salt and Hollick, 1944; Raw, 1955). Present-day health and safety regulations would preclude the use of such solvents but I have no knowledge of safe alternatives.

A precision flotation system, capable of dealing with smaller samples, uses the Griffiths-Thind flotation flask (Thind and Griffiths, 1979), subsequently refined by Thind and Wallace (1984), and now extensively used to extract mites from dust and food samples. This apparatus would be most useful for specific cases of soil extraction, particularly where initial concentration has been carried out and very small specimens are involved.

Collection from Marine and Freshwater Sediments

Aquatic mites (*Hydrachidia*) are collected in similar fashion to other aquatic arthropods. Bottom dredge samples should be washed through suitable meshed filters, or, if mainly composed of sediment, filters may not be necessary. The final sample must be examined in a white porcelain or plastic tray, where they can be easily seen and picked out using fine-diameter pipettes.

Benthic mites are best collected using inverted traps as invented by Pieczynski (1961), later improved upon by Conroy (1971). Bottom mud can be treated by placing the sample in concentrated magnesium sulphate solution. After about 20 min, because of the high specific gravity, mites rise to the surface, from which they can be collected (Efford, 1965). Concentrated samples would be suitable for extraction in the Griffiths-Thind flask.

Bottom sediments should be reduced in volume using sedimentation differentiation and screening techniques. A final removal of silt can be achieved by using a 100-105 μm screen, yielding a range of specimens from a few microns up to about 100 μm (Newell, 1971).

Coral and Algae

Collecting mites from such marine material requires the sample to be vigorously agitated, whilst at the same time being broken down into smaller pieces. In practice, a 10-15 litre container of seawater, to which 15-20 ml of chloroform is added after agitation begins, is sufficient to treat a reasonably sized sample. The anaesthetic, once dispersed, causes the mites to loose their hold on the substrate.

A coarse-meshed inner liner should be fitted inside the container, which allows the larger pieces of debris to be retained when the liner is lifted from the container. In the process of lifting, the debris in the liner should be washed with a jet of clean seawater.

The final sediment is collected in a fine-mesh bag, securely fastened and transferred to a holding vessel containing a suitable preserving fluid. Newell (1971) included about 2 ml rose bengal dye per 100 ml preservative, which gives halacarid mites a distinctive red coloration, making it much easier to pick them out.

Final processing is carried out by using a series of graded sieves over which a controlled jet of water is directed. Nylon meshes in a very wide range of sizes down to a few microns aperture are commercially available.

Processing the Extraction Fraction

If live material is required, the collecting bottle attached to the funnel stem should contain water or a moistened substrate such as filter paper. Individual mites can then be picked out for observation or culturing. If the material is to be preserved, a solution of 70% (v/v) alcohol should be used in the bottle. Up to 5% (v/v) glycerol may be added to prevent the specimens drying out during storage. Properly sealed, this preparation can be stored for long periods. However, for museum material it is better to use a narrow-necked pipette to transfer the mites to small tubes (12-25 mm long and 6 mm diam). The tube is plugged with a porous cork made of plastic, botanical pith or polyporus (Hobart, 1956). The small tube, together with a collection label, is placed in a slightly larger tube, which is filled with alcohol, plugged with cotton wool and inverted with other tubes into a museum glass jar filled with alcohol and sealed with a ground-glass stopper.

A simple technique, used by Michael (1884), for obtaining relaxed specimens is to collect the mites alive, and then pour a small quantity of boiling water into the collecting vessel. This provides a specimen in which the appendages, including the mouthparts, are well extended. Such specimens can then be placed in a preservative liquid, or examined on a slide.

There are a number of simple instruments for handling mites. The choice will depend on the size of the mite and the medium in which they are contained. Needles can be made from a No. 0 or No. 1 size entomological pin, fixed into a suitable holder. A lifting spoon can be made by flattening the tip of a dissecting needle or pin. The neck of the spoon needs to be bent to an angle of about 80°. A third essential instrument is a No. 00 sable paint brush. For the smaller acari it will be necessary to remove most of the hairs, even from the finest brush.

The use of polyvinol-alcohol (PVA), first proposed by Heinz (1952), subsequently modified by Boudreaux and Dosse (1963), both clears and provides a permanent mount for all light to medium sclerotized mites. The mountant can be hardened before ringing by gently heating on a slide warmer at about 50°C overnight. It is important to use the correct grade of PVA, one with a low percentage hydrolysis and low viscosity, the latter measured in 'centipoises'. The strengths are usually displayed after the trade name of the PVA as two figures '52-22', which figures represent concentrations suitable for mite slides. Because this is now such an important technique for obtaining permanent mite mounts, the recipe and ingredients are given below. Mounted specimens may be recovered by soaking the slide in warm lactophenol.

Very heavily sclerotized mites (the *Cryptostigmata*) may require extra clearing before a permanent mount is made. A solution of 70-80% (v/v) lactic acid, a few drops placed in a cavity slide and then warmed over night or longer if necessary, will usually suffice. The specimen needs to be washed in a drop of PVA to remove the lactic acid before being permanently mounted.

Heinz Mounting Medium

Polyvinyl alcohol (PVA)	10 g
Distilled water	40-60 ml
Glycerol	10 ml
Phenol solution (1.5% v/v)	25 ml
Lactic acid (85-92% v/v)	35 ml

For all but the largest specimens it is recommended that coverslips no larger than a diameter of 12 mm be used, and no more than two or three specimens be placed on any one slide. The most successful way to mount mites is to take a bottle cork (smallest diameter 30 mm and depth of about 20 mm) then darken the smaller surface of the cork with black ink. To prepare the slide place a 12 mm cover slip onto the dark surface, then with a fine needle apply one drop of PVA to its centre. Place the cork under the binocular microscope, add and orientate the specimens, then gently lower a clean slide down onto the cover slip so that it is centred on the slide. With practice, the correct size drop can be achieved which will give an even spread of mountant beneath the cover slip.

Add water to PVA powder, stirring constantly, the mixture being heated in a water bath at just below boiling-point. Add lactic acid and stir for a few minutes. Add glycerol and stir until smooth. Cool until lukewarm, then add the chloral hydrate, which has been previously dissolved in the phenol solution. Store in a brown glass bottle.

Bibliography

Berlese, A. (1905) Apparecchio per raccogliere presto ed in gran numero piccoli atropodi. *Redia* 2, 85.

Boudreaux, H.B. and Dosse, G. (1963) The usefulness of new taxonomic characters in females of the genus *Tetranychus* Dufour (*Acari:Tetranychidae*). *Acarologia* 5, 14-33.

Conroy, J.C. (1971) A new method for trapping water mites in the benthos of a lake. *Proceedings of the 3rd International Congress of Acarology, Prague*, pp. 151-157.

Efford, I.E. (1965) The ecology of the water mite *Feltria romijni* Besseling. *Journal of Animal Ecology* 34 , 233-251.

Evans, G.O., Sheals, J.G. and Macfarlane, D. (1961) *The Terrestrial Acari of the British Isles,* Vol. I. *Introduction and Biology.* British Museum (Natural History), London.

Haarlov, N. (1947) A new modification of the Tüllgren apparatus. *Journal of Animal Ecology* 16, 115-121.

Heinz, K. (1952) Polyvinylalcohol Lactophenol Gemisch als Einbettungsmittel für Blattläuse. *Naturwissenschaften* 39, 285-286.

Hobart, J. (1956) The use of polyporus for plugging small vials. *Entomological Monograph Magazine* 92, 277-278.

Macfadyen, A. (1953) Notes on methods for the extraction of small soil arthropods. *Journal of Animal Ecology* 22, 171-182.

Michael, A.D. (1884) *British Oribatidae,* Vol. 1. Ray Society, London.

Murphy, P.W. (1955) Note on processes used in sampling, extraction and assessment of the meiofauna of heathland. In: Kevan, D.K.McE. (ed.) *Soil Zoology.* Butterworths, London. pp. 338-340.

Newell, I.M. (1971) *Halacaridia* (*Acari*) collected during Cruise 17 of the R/V Anton Bruun, in the southeastern Pacific Ocean. *Anton Bruun Report* 8, 3-58.

Pieczynski, E. (1961) The trap method of capturing water mites (Hydracarina). *Ecological Pollution Bulletin* 7, 111-115.

Raw. F. (1955) A flotation extraction process for soil micro-arthropods. In: Kevan, D.K.McE. (ed.) *Soil Zoology.* Butterworths, London, pp. 341-346.

Salt, G. and Hollick, F.S.J. (1944) Studies of wireworm populations. 1. A census of wireworms in pasture. *Annals of Applied Biology* 31, 52-64.

Thind, B.B. and Griffiths, D.A. (1979) Floatation technique for the quantitative determination of mite populations in powdered and compacted foodstuffs. *Journal of the Association of Officers of Analytical Chemistry* 62, 278-282.

Thind, B.B. and Wallace, D.J. (1984) Modified floatation technique for the quantitative determination of mite populations in feedstuffs. *Journal of the Association of Officers of Analytical Chemistry* 67, 866-868.

Tüllgren, A. (1918) Ein sehr einfacher Auslesgeapparat für terricole Tierformen. *Zeitschrift für Angewante Entomologie* 4, 149-150.

Meiofauna in Marine and Freshwater Sediments 15

MAGDA VINCX

Universiteit Gent, Department of Morphology, Systematics and Ecology, Marine Biology Section, K.L. Ledeganckstraat 35, B-9000 Gent, Belgium.

The Role of Meiofauna in Marine and Freshwater Ecosystems

The meiofauna is defined on a methodological basis as all metazoans retained on a sieve of 42 μm (Mare, 1942). Meiofauna occur in freshwater and marine habitats, although most ecological studies on meiofauna have been performed in the marine environment. Meiofauna occur from the splash zone on the beach to the deepest sediments in the sea: they are found in all types of sediments (clay to gravel), are common as epiphytes on seagrasses and algae, in sea ice and in various animal structures (as commensals or parasites). Only members of the free-living interstitial community which live between sediment grains will be described in this chapter.

Twenty-three higher taxa of the thirty-three metazoan phyla have some meiobenthic representatives: *Nematoda, Turbellaria, Oligochaeta, Polychaeta, Copepoda, Ostracoda, Mystacocarida, Halacaroidea, Hydrozoa, Nemertina, Entoprocta, Gastropoda, Aplacophora, Brachiopoda, Holothuroidea, Tunicata, Priapulida, Sipunculida.* The phyla *Gastrotricha, Gnathostomulida, Kinorhyncha, Loricifera* and *Tardigrada* are exclusively meiobenthic (meiobenthos = meiofauna living in sediments). Nematodes, copepods and turbellarians comprise more than 95% of the meiofauna in most sediments. A comprehensive guide on sampling procedures and ecology of the meiofauna is presented in Higgins and Thiel (1988) and in Giere (1993). In this chapter, a general methodology about meiofauna is presented first and then nematodes, copepods, kinorhynchs, gastrotrichs, gnathostomulids, nemertineans, tardigrades and oligochaetes are discussed briefly.

In the last 20 years, much ecological information about meiofauna is illustrating more and more the picture that these small creatures are important in marine sediments, both indirectly, by processes such as bioturbation and the stimulation of bacterial metabolism, and also directly, as food sources for organisms in higher trophic levels, such as shrimp and juvenile fish. They also play an important role in pollution research.

On average, 1-10 million individuals of meiofauna are present in 1 m^2 sediment, which represents, however, a biomass of only a few grams. These abundance:biomass ratios vary according to temperature, water depth, tidal exposure, grain size, etc. Meiofaunal productivity, however, is very high with P/B ratios ranging from 6 to 50 (Vranken *et al.*, 1986).

In almost all meiobenthic studies, the majority of the fauna has been found in the upper 2 cm of sediment. Vertical zonation is typically controlled by the depth of the redox potential discontinuity (RPD) layer, i.e. the boundary between aerobic and anaerobic sediments. The primary factor or 'super parameter' responsible for vertical gradients in the RPD is oxygen, which determines the redox potential, as well as the oxidation state of sulphur and various nutrients. When redox potentials drop below +200 mV (measured with microelectrodes, and not always directly by the colour change from the grey-brown of the upper sediment layer to the black of the deeper layers), metazoan meiofauna densities greatly decrease (McLachlan, 1978). Nevertheless, numerous geochemical cycles and microbial processes together with the bioturbative impact of sediment fauna create a complicated three-dimensional 'landscape of the redox potential' (Ott and Novak, 1989) with oxic/anoxic areas often surrounding a vertical redox threshold. The sulphidic ecosystem, dominated by the presence of reduced substances such as dissolved sulphide, methane and ammonium, is much too complex to allow for simple answers to the previous debate about the existence of a 'thiobios' (Giere, 1993 and references therein). Typical representatives of the reduced redox potential layers are specialized nematodes, oligochaetes and turbellarians. Sediment grain size is another key factor affecting the structure of meiobenthic communities (density, biomass and diversity).

Sampling/Extraction Techniques

Sampling strategy has to be adapted to the kind of sediment being examined (Kramer *et al.*, 1992), although we aim at one sampling technique for several types of sediments in order to work with a standardized technique for biodiversity comparison between different regions. Baseline studies are missing for most of the habitats still to be investigated for biodiversity studies, and, therefore, one standardized method is presented in this chapter.

Almost all investigators have considered the problems associated with sampling, and many have arrived at individual solutions to their problems. The more common sampling methods and equipment for use in a wide variety of habitats are described here.

Biodiversity measurements are only relevant if they can be calculated from samples taken on a quantitative basis and so only quantitative sampling methods are discussed (of course, all processing methods are suitable for qualitative methods as well).

Sampling methods, the number of samples, fixation and extraction techniques are mostly the same for all meiofauna groups (certainly for the hard-bodied taxa such as nematodes, copepods, etc.). Microscopical examination and identification will be discussed for the dominant taxa separately.

Sampling methods

Methods for meiofauna in general, as well as for all the individual taxa, have been discussed in detail in Higgins and Thiel (1988).

Coring

In sediments, coring is the best quantitative sampling technique. Corers are devices of known sampling area. Most are cylindrical and can be made from tubing or piping of any rigid material locally available. The diameter chosen depends on the volume and depth of the sample required and whether the reduced layers are being sampled or not. Corers with an inner diameter of 2 cm have been used in many habitats and provide a meiofaunal sample that can be sorted in its entirety. Smaller corers (1 cm diam. or less) are desirable in sediments where densities are usually high, or to determine small-scale distributions. **For biodiversity studies, corers of 2-4 cm diam. up to at least 5 cm sediment depth are recommended.** Kramer *et al.* (1992) mentioned that there is no optimum between surface area (corer diam.) and number of organisms per sample. In practice, experienced scientists have arrived at a procedure that involves somewhat larger size samples in sandy sediments and smaller ones in muddy sediments. However, for the most dominant phyla of the meiofauna, the nematodes, it has been shown by several authors that a standard corer surface of about 10 cm^2 is appropriate for all types of sediment.

In intertidal areas, tubes with the characteristics described above may serve as a sampler for the upper portion of the sediment. A tight-fitting stopper secured in the upper end of the coring tube or the piston of the syringe will provide suction to hold the sediment in place while the corer is removed. Another stopper may secure the lower end of the tube for transport.

In deeper waters, the different types of box corers or multiple corers are to be preferred to grabs (Bett *et al.*, 1995). An ideal corer should penetrate the sediment without a shock wave and as slowly as possible. Out of the box corers, similar sampling corers can be used as for the intertidal conditions.

Number of Samples

The number of samples (as well as sample size) depends on the problem being studied. Variability in density of meiofauna appears to be mostly on a scale of a few centimetres and again on a scale of many km, or even more, depending on the substratum heterogeneity. In studies, aiming at studying small-scale horizontal distribution obviously the largest amount of the smallest possible samples is required, with the important restriction that sample size must be large compared with the individuals being sampled. When the problem consists, as is usually the case, in obtaining an estimate of meiofaunal density and diversity, there may be two alternative solutions. The first is to destroy all small-scale variability in the sample and take a sample as big as possible (e.g. mixing three corers), mix it thoroughly and, if necessary, take subsamples for further analysis. Second, when the intrinsic pattern is not destroyed, aggregation will require that the sample size is as small as possible and the number of samples taken as large as possible. Most statistical analyses are robust when they are based on a large number of error degrees of freedom (Green, 1979) and so, for 'general' biodiversity studies, it is recommended that three replicates of 10 cm^2 are analysed separately.

Fixation

Sediment samples should be stored, after addition of 4% (v/v) neutral formalin, in warm (60°C) seawater solution, in polyethylene bottles prior to analysis. A warm solution of formalin is advised to prevent nematodes rolling up, which will make identification nearly impossible. Formalin may be neutralized with a saturated solution of $LiCO_3$, taking care that no excess buffering solution is added. (Excess $LiCO_3$ causes crystallization of formalin around the animals, which makes them impossible to identify.) Preserved samples may be stored for many years until analysis.

Extraction Techniques

Decantation

The extraction of meiofauna from sediments is easy when the sediment is a sand with low amounts of detritus or silt-clay. Simple decantation on a sieve is often satisfactory and a 38-42 µm sieve is standard for meiofauna work. For decantation:

- put the sample in a vial of 5 litres and stir well
- put the supernatants on a sieve of 38 µm diameter
- repeat this action 10 times on the same sieve.

Density gradient centrifugation

The extraction from mud or detritus (after the sand has been removed by decantation or other methods) is done most efficiently using a density gradient in a centrifugation procedure (Heip *et al.*, 1974, 1985). Three possible liquids with a density larger than the density of meiofauna (i.e. 1.08 for nematodes, but this seems to be workable for all hard-bodied meiofauna) can be used, e.g. 'Ludox HS40', 'Ludox AS' and sugar. 'Ludox' is a silicasol (a colloidal solution of SiO_2), which causes no plasmolysis. 'Ludox HS40' is toxic and so may be used for fixed material. 'Ludox AS' (the most expensive) is not toxic and can be used when living meiofauna has to be separated from sediments. For both types of 'Ludox', a 50% solution in distilled water is used (density of 1.15). A sugar solution is made by adding sugar to 700 ml hot water, until 1000 ml is reached, and is best used when sediments contain a lot of organic detritus (e.g. in salt marsh and mangrove sediments). The method developed in our laboratory consists in the slightly modified procedure of Heip *et al.* (1985).

- Rinse the fixed sample thoroughly with tap water, to prevent flocculation of 'Ludox', over a sieve of 38 μm. Decantation can be done comparable to the sandy sediment sample.
- Bring the sample from the sieve in a centrifugation tube as large as available
- Add the 'Ludox' solution (60% 'Ludox' and 40% water; density = 1.18).
- Centrifuge at 1800 g in water for 10 min.
- Repeat the centrifugation three times more.

The supernatant is finally rinsed over a 38 μm sieve for some time with water, because 'Ludox' and formalin react and form a gel which is difficult to wash out. Centrifugation with sugar only takes 5 min, but requires washing with much water afterwards, because the sugar solution causes plasmolysis in nematodes.

After extraction, 4% neutral formalin is added again to the treated sample. Counting is facilitated by staining of the entire sample with rose bengal (1% for 48 h). Preserved samples can be stored until analysis.

Counting

Using a stereomicroscope and a counting box with a grid of 10 × 5 or 10 × 10, the densities of animals are determined either by counting all specimens or by taking out only the first 200 animals of a specific group encountered (e.g. the first 200 nematodes, which will be identified to species level and, therefore, mounted in glycerol).

Microscopic Examination and Identification

Nematodes

After fixation, nematodes must be transferred to anhydrous glycerol. Specimens are transferred from formalin to glycerol through a series of ethanol-glycerol solutions to prevent the animals from collapsing. They are picked out and put into a cavity block (recipient) under a stereoscopic microscope into a solution of 99% formalin (4%) and 1% glycerol. The recipient is then put into a vial containing a bottom of 95% (v/v) ethanol at 35°C for about 12 h.

At 35°C, ethanol is evaporated into the solution of formalin and glycerol. After 12 h (e.g. overnight), the cavity block (with nematodes) is partly covered and put in an oven at 35°C. Every 2 h, some drops of a solution of ethanol with glycerol (95%:5% v/v) is added with a pipette. After about 6 h, some drops of an ethanol and glycerol (50%:50% v/v) solution are added. The cavity block stays partly open at 35°C until all ethanol is evaporated and the nematodes remain in pure glycerol.

Animals may be mounted on glass slides when in glycerol. For this a paraffin ring is put on a slide, within a small droplet of glycerol and 5-10 nematodes of about the same thickness are put into the glycerol drop. A cover glass is put on the droplet and slightly heated at 40°C in order to let the paraffin melt. For permanent slides, the cover glass may be sealed with 'Glyceel', 'Clearseal' or 'Bioseal' (certainly necessary in tropical areas).

Whole preparations are usually satisfactory for species identification. A good-quality microscope with a ×100 oil-immersion lens is required. Comprehensive guides for the identification of nematodes have been available for 10 years, and pictorial keys to genus level are especially useful (e.g. Platt and Warwick, 1983).

Copepods

Sorted specimens should be placed in 70% (v/v) ethanol if prolonged storage is anticipated. Initial identification to species level almost always involves examination of the appendages and setation, and may required observations of the head appendages and body surface spinulation patterns. Such details are seen most clearly in dissected specimens mounted on slides. Once species have been identified, they can be recognized subsequently on gross structures visible with a stereomicroscope. For permanent mounts, whole animals, or dissected parts, should be mounted in a gum-arabic-based medium, or in fluid mountants, such as glycerol or lactic acid. Lactic acid is a powerful clearing agent and is inadvisable for long-term storage, but this problem is substantially reduced if it is mixed with glycerol (at about 1 part of lactic acid in 4 of glycerol). Polyvinyl lactophenol has been widely used, but tends to shrink with time and there is little doubt that the gum arabic mountants give the best long-term results (Wells, 1988).

Kinorhynchs

Sorted specimens should be placed in 70% (v/v) ethanol, if prolonged storage is anticipated. The identification of kinorhynchs demands that the specimens be mounted in a perfectly dorsoventral aspect; specimens mounted otherwise are virtually worthless. Specimens should be moderately cleared in order to observe the details of external morphology (use 62.5% v/v chloral hydrate solution) (Hoyer's medium). Carefully sealing the cover slip is extremely important in producing a permanent mount (Higgins, 1988).

Turbellarians

(See this volume, Chapter 17, for details of this group.)

Gastrotrichs, gnathostomulids and nemertineans

Since gastrotrichs are soft-bodied meiofauna, identification and extraction methods are the same as those described for turbellarians (see this volume, Chapter 17). Since many gastrotrichs are extremely small, only small sediment quantities should be used when applying the $MgCl_2$-decantation method. A fine mesh size of 35 μm is recommended and an Erwin loop is better than a pipette to pick them out. Gnathostomulids are mainly found in larger quantities of sediment.

Tardigrades

Animals are transferred with a drop of 2% (v/v) formalin to microslides and covered with a cover slip (Kristensen and Higgins, 1984). The formalin preparation is infused with a 10% (v/v) solution of glycerol in 96% (v/v) ethyl alcohol and allowed to evaporate to glycerol over a period of several days. The resulting whole mount can be sealed with 'Clearseal'.

Oligochaetes

Identification of aquatic oligochaetes is, in most cases, possible only if sexually mature specimens are available. Most of the taxonomically important features are restricted to the genitalia and a few other internal structures. For permanent mounts, Canada balsam or artificial balsam can be used as a mounting medium. For temporary mounts of fixed animals, 4% formalin:glycerol is an excellent medium as it has a clearing effect, but does not harden the specimen. The use of lactophenol for clearing/mounting should be restricted to studies where deterioration of the material after some time (several months) is permissible, and where the species studied have enough setal characteristics, such as cuticular penis sheaths, to ensure that identification is possible (provided always that taxonomy of the species studied has already been well established). Soft internal structures are not always visible in lactophenol (Erseus, 1988; see also this volume, Chapter 20).

Bibliography

Bett, B.J., Vanreusel, A., Vincx, M., Soltwedel, T., Pfannkuche, O., Lambshead, P.J.D., Gooday, A.J., Ferrero, T. and Dinet, A. (1995) Sampler bias in the quantitative study of deep-sea meiobenthos. *Marine Ecology Progress Series* 104, 197-203.

Erseus, C. (1988) Oligochaeta. In: Higgins, R.P. and Thiel, H. (eds) *Introduction to the Study of Meiofauna.* Smithsonian Institution Press, Washington DC, pp. 349-354.

Giere, O. (1993) *Meiobenthology. The Microscopic Fauna in Aquatic Sediments.* Springer Verlag, Berlin.

Green, R.H. (1979). *Sampling Design and Statistical Methods for Environmental Biologists.* J. Wiley, New York.

Heip, C., Smol, N. and Hautekiet, W. (1974) A rapid method of extracting meiobenthic nematodes and copepods from mud and detritus. *Marine Biology* 28, 79-81.

Heip, C., Vincx, M. and Vranken, G. (1985) The ecology of marine nematodes. *Oceanographic Marine Biology Annual Review* 23, 399-489.

Higgins, R.P. (1988) Kinorhyncha. In: Higgins, R.P. and Thiel, H. (eds) *Introduction to the Study of Meiofauna.* Smithsonian Institution Press, Washington DC, pp. 328-331.

Higgins, R.P. and Thiel, H. (1988) *Introduction to the Study of Meiofauna.* Smithsonian Institution Press, Washington DC.

Kramer, K.J.M., Warwick, R.M. and Brockmann, U.H. (1992) *Manual of Sampling and Analytical Procedures for Tidal Estuaries.* TNO Institute of Environmental Sciences, Den Helder.

Kristensen, R.M. and Higgins, R.P. (1984) Revision of *Styraconyx* (Tardigrada/Halechiniscidae), with descriptions of two new species from Disko Bay, West Greenland. *Smithsonian Contributions to Zoology* 361, 1-40.

McLachlan, A. (1978) A quantitative analysis of the meiofauna and the chemistry of the redox potential discontinuity zone in a sheltered sandy beach. *Estuarine and Coastal Marine Science* 7, 275-290.

Mare, M.F. (1942) A study of a marine benthic community with special reference to the micro-organisms. *Journal of the Marine Biological Assocaiation of the United Kingdom* 25, 517-554.

Ott, J. and Novak, R. (1989) Living at an interface: meiofauna at the oxygen/sulfide boundary of marine sediments. In: Ryland, J.S. and Tyler, P.A. (eds) *Reproduction, Genetics and Distribution of Marine Organisms.* Olsen & Olsen, Fredensborg, pp. 415-422.

Platt, H.M. and Warwick, R.M. (1983) Freeliving marine nematodes. Part I. British Enoplids. *Synopses of the British Fauna (New Series)* 28.

Platt, H.M. and Warwick, R.M. (1988) A synopsis of the freeliving marine nematodes. Part II. British Chromadorids. *Synopses of the British Fauna (New Series)* 38.

Vranken, G., Herman, P.M.J., Vincx, M. and Heip, C. (1986) A re-evaluation of marine nematode productivity. *Hydrobiologia* 135, 193-196.

Wells, J.B.J. (1988) Copepoda. In: Higgins, R.P. and Thiel, H. (1988) *Introduction to the Study of Meiofauna.* Smithsonian Institution Press, Washington DC, pp. 380-388.

Rotifers 16

Parke A. Rublee

Biology Department, University of North Carolina at Greensboro, Greensboro, NC 27412-5001, USA.

The Importance of Rotifers in Soils and Sediments

Rotifers (phylum *Rotifera*) are predominantly aquatic animals although they are also able to survive in thin films of water associated with terrestrial surfaces. They are found in marine and freshwater sediments, associated with moss and liverwort surfaces, and in moist to saturated terrestrial soils. In marine sediments they are placed in the group known as meiofauna, generally known as those benthic organisms between about 31 and 1000 μm in size. In terrestrial soils, they are included in the group known as mesofauna, with a size range from about 200 μm to 10 mm, although this size range may exclude some of the smaller rotifers. These terms will be used interchangeably here.

Research on rotifer abundance, diversity and importance has focused predominantly on planktonic freshwater systems where they may be the most abundant suspended zooplankton. In these systems they play important roles in microbial food webs

- as consumers of algal and bacterial primary and secondary production,
- as an important food resource for larger animals, and
- by mobilizing and immobilizing important nutrients that affect algal and bacterial production.

Rotifers attached to surfaces and in soils and sediments undoubtedly play similar roles, although they have not received the same level of attention as planktonic forms and their importance in soils and sediments is less well understood. Several excellent recent reviews note that this is an area in need of study (Wallace and Snell, 1991; Nogrady *et al.*, 1993).

Despite limited studies, common patterns have emerged with regard to rotifers on or within sediments and soils. First, rotifers are common inhabitants of sediments and soils (Table 16.1), and they have a worldwide distribution.

Regardless of location, abundance tends to be highest in oxygenated surface layers and decreases with depth in both soils and sediments. Second, although rotifers are subject to the usual physical, chemical and biological factors which affect organism distribution, there are two factors that are critical: availability of water and adequate sediment or soil structure. Rotifers are essentially aquatic animals and generally feed on small particles via an oral ciliary apparatus. Thus, a thin film of water is necessary for metabolically active populations. However, a constant supply of water is not necessary, and rotifers may persist through dry periods in resting states or cysts, a process known as anhydrobiosis and common to soil rotifers. Additionally, the sediment structure must be coarse enough to contain interstitial spaces. If the silt and clay content of the soil or sediment is too high, it may preclude the presence of rotifers.

A third emerging pattern is that rotifers are integral components of microbial food webs in soils and sediments. They are involved in nutrient cycling, are important grazers of small particles and may, in turn, be consumed by larger grazers (Fig. 16.1).

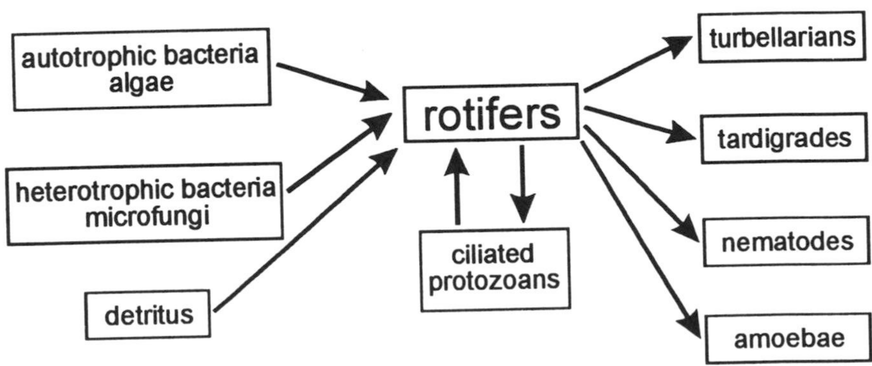

Fig. 16.1. Generalized soil or sediment food web. Only interactions between rotifers and other soil organisms are shown.

Their quantitative importance within these food webs appears to vary, however. In marine sediments, fewer species are represented, and they are probably not of major importance, although they may reach high abundance in the littoral benthos and in marsh sediments. In freshwater sediments and terrestrial soils, they may be of much greater significance. For example, Palmer (1990) found that rotifers comprised 35-85% of the individuals of the fauna in

Table 16.1. Selected values for rotifer density in soils and sediments obtained from literature.

Biotope	Depth	Abundance	Extraction method	Reference
Terrestrial soils				
shortgrass prairie (USA:CO)	top 8 cm	$3.2\text{-}9.54 \times 10^4$ m^{-2}	heated funnel	Andersen *et al.* (1984)
meadows and fields	?	3.0×10^5 m^{-2}	-	Franz (1950) in Pourriot (1979)
tropical forests and prairie	?	$3.0\text{-}45.0 \times 10^4$ m^{-2}	-	Maldague (1959, in Pourriot
(1979)				
pine forest, Sweden	top 10 cm	$20\text{-}120 \times 10^4$ m^{-2}	heated funnel	Sohlenius (1979)
pine forest, Sweden	top 10 cm	$0.8\text{-}22.4 \times 10^5$ m^{-2}	heated funnel	Sohlenius (1982)
Freshwater sediments				
sandy creek bottom (USA:OH)	top 2 cm	0-8.6 cm^{-2}	serial decantation with $MgCl_2$	Evans (1984)
pond sediments (USA:SC)	top 7-10 cm	2.7-74.0 cm^{-2}	elutriation	Oden (1979)
arctic river (USA:AL)	rock surface	0-0.84 cm^{-2}	rinsing	Rublee and Partusch-Talley (1995)
gravel brook (Austria)	top 1 cm	0-0.15 cm^{-2} 0-46 l^{-1}	Hess sampler for bed surface standpipes for interstitial water	Schmid-Araya (1993)
sandy beach	?	up to 23 cm^{-2}	?	Evans (1982) in Turner (1988)
freshwater sand beaches	?	0-141 cm^{-2}	various	various studies in Pennak (1988)
Marine/estuarine sediments				
estuarine sediments (Denmark)	top 4 cm	0-24 cm^{-2}	seawater ice	Thane-Fenchel (1968)
intertidal sandy beach (Germany)	?	1.3-4.5 cm^{-3}	seawater ice	Tzschaschel (1980)

the area of subsurface flow (hyporheic zone) of a stream. The true importance of rotifers in soil food webs may be difficult to judge, however, since the presence of resting stages and cysts and reproduction by parthenogenesis mean that population sizes can increase by orders of magnitude rapidly over short time intervals under favourable conditions. This may also involve changes in population structure, thus affecting diversity measures. Pourriot (1979) has suggested that rotifers can be used as indicator species for soils and sediment systems, but much additional work needs to be done to establish firmly the niches for various species.

Taxonomic Considerations

The phylum *Rotifera* includes about 2000 species, most of which are found in freshwater environments. (Some European workers tend to rank rotifers as a class within the phylum *Aschelminthes*, rather than as a separate phylum.) The taxonomy of rotifers is based on a number of morphological characteristics of this diverse group. While the diversity of form is valuable for identification to the generic level, considerable variability and morphological plasticity within genera make species identification more problematic. In some genera, species identification is based on the structure of the trophi, the chitinous jaws within the mastax, the muscular pharynx. These difficulties suggest a need for revision of the group (Nogrady *et al.,* 1993). Rotifers in soils and sediments come predominantly from two orders, the *Bdelloidea* (class *Digononta*) and the *Ploimida* (class *Monogononta*). In marine benthic habitats, representative genera of the *Ploimida* comprise the vast majority of species encountered, but, during the transition from littoral zones to freshwater sediments and finally to terrestrial soils and mosses, the abundance of bdelloid taxa increase to > 90%.

The need for a revision of the phylum *Rotifera* presents a difficulty in biodiversity studies. One may still identify and classify according to the best keys available, but it should be understood that taxonomic revisions may alter diversity estimates. The advent of molecular biology techniques, specifically the use of small subunit ribosomal gene sequences as a phylogenetic tool and gene probes for taxonomic identification, would appear to be a very desirable aid in development of a clear taxonomy and a potentially powerful tool for identifying rotifers.

Methods for the Extraction of Rotifers

Methods for quantification of rotifers on or within substrates have generally derived from those used for meiofauna. Unfortunately, in some approaches soft-bodied rotifers, especially bdelloids, may be damaged or not recovered quantitatively by these methods, and the investigator should consider some preliminary testing of methods to determine which is best suited for the site to be studied. In some cases, however, the methods generally used for meiofauna or mesofauna are suitable and rotifers are simply evaluated in the same sample as nematodes.

Sample Collection

Collection of mosses, humus or other plant material to study epiphytic rotifers is usually done by transfer of harvested plant material to a container for transport to the laboratory. In the laboratory, the sample is agitated vigorously in water, the plant material removed, and fauna remaining in the water can then identified and enumerated. Rublee and Partusch-Talley (1995) have used a related method of gently rinsing rock surfaces with spray from a squeeze bottle to determine rotifer abundance in Arctic streams. Peters *et al.* (1993) note that there is incomplete removal of rotifers by such methods, and they describe a quantitative procedure for organisms that can be removed by agitation. Briefly, the extraction (agitation) is repeated several times in clean water. Cumulative plots of individuals extracted vs the number of extractions should approach an asymptote representing the maximum number of individuals that can be extracted by this method. The asymptotic limit is easily calculated by a regression of the reciprocals of these values, where the y-intercept represents the reciprocal of the limit. The resultant values should be reported as numbers per dry weight or surface area of plant material to facilitate comparison with other studies.

Collection of sample material in sediments or saturated soils is usually done with piston or box corers or dredges (see Fleeger *et al.*, 1988). Care must be taken to minimize the disturbance of sediment material and interstitial water in order to prevent damage, movement or loss of organisms in the sample due to flushing or squeezing. Soil samples from terrestrial sites may also be collected by coring devices or by simply digging a known volume. In general, samples should be returned to the laboratory for extraction of rotifers within a few hours, and may be kept on ice during this interval.

Extraction and Observation

Once soil or sediment samples are collected, organisms may be identified and enumerated by direct observation of the samples or by extraction of organisms from the soil or sediment, followed by direct microscopic observation of the living material. Several methods have been used to extract rotifers from soil or sediment material, including decantation, funnel filtration, elutriation and flotation. The activity of live organisms may be reduced by the addition of viscous agents such as methyl cellulose (2% w/v aqueous solution). Some investigators have gone through additional steps of relaxation (anaesthetization) and fixation prior to the enumeration and identification steps, and, in some cases, prior to extraction. Visualization of the trophi is accomplished by adding a small amount of bleach (5% (v/v) sodium hypochlorite) to dissolve soft tissues of rotifers placed in a depression slide (Myers, 1937; Edmondson, 1959).

Direct observation

This is the simplest, but perhaps most tedious approach to enumerating rotifers. Aliquots of the soil or sediment are placed in a small dish or trough (e.g. 1 g soil/sediment in a Petri dish) with enough filtered water of the appropriate salinity to distribute the sample across the bottom of the dish. It is then carefully scanned with a dissecting microscope for living organisms, which are removed by capillary pipettes to separate dishes for identification and enumeration at higher magnification with a compound microscope. Isolated samples may be maintained for periods of several weeks if the water is changed daily (Pourriot, 1979). It has also been suggested that incubating dry soil samples with added water for several weeks will allow resting stages and eggs to hatch (May, 1986).

Decantation

This method simply involves disturbing or agitating the sample in a volume of water, waiting for a short period to allow for settling of sediment particles, pouring off the supernatant, and repeating the process. The supernatant material is combined and allowed to settle or passed through a filter. Collected organisms are then identified and enumerated. The use of anaesthetics or fixatives may be combined with decantation to improve recovery. For example, Evans (1984) utilized a concentration of 1% (w/v) $MgCl_2$ during decantation of stream sediments.

Funnel filtration

In this method, the sample is flushed with water to wash out rotifers or other meiofaunal organisms. 'Funnel' methods generally utilize gravity flushing of water through samples. In the simplest form, the soil or sediment sample is placed over gauze or netting which lines a funnel suspended over a collection basin (Fig. 16.2A). Water is added to the surface of the sediment sample and organisms are collected in the wash water. While this method may work well for many rotifers, it may not provide a complete extraction for those with adhesive glands or other attachment mechanisms.

Many modifications of the method have been used to improve efficiency of extraction, but two should be mentioned. The first holds the water over the sample for a period of 1 or more hours, during which the water is heated up to 29-30°C by suspending a light bulb over the funnel. The water is then allowed to flush through the funnel (Overgaard-Nielsen, 1948; Sohlenius, 1979). The second method, developed for marine sediments, is the 'seawater ice' method of Uhlig (1964). In this method, seawater is frozen and placed on top of the sample

(Fig. 16.2B). As the ice thaws, organisms are subjected to a flush of cold water, which is at first very saline and becomes progressively less saline, although the organisms are finally collected in water of ambient salinity. Both of these methods cause relaxation of the organisms, reducing adherence to the substrate and thereby allow them to be flushed out of the sample. However, efficient extraction depends on a high degree of porosity of the sediment or soil sample.

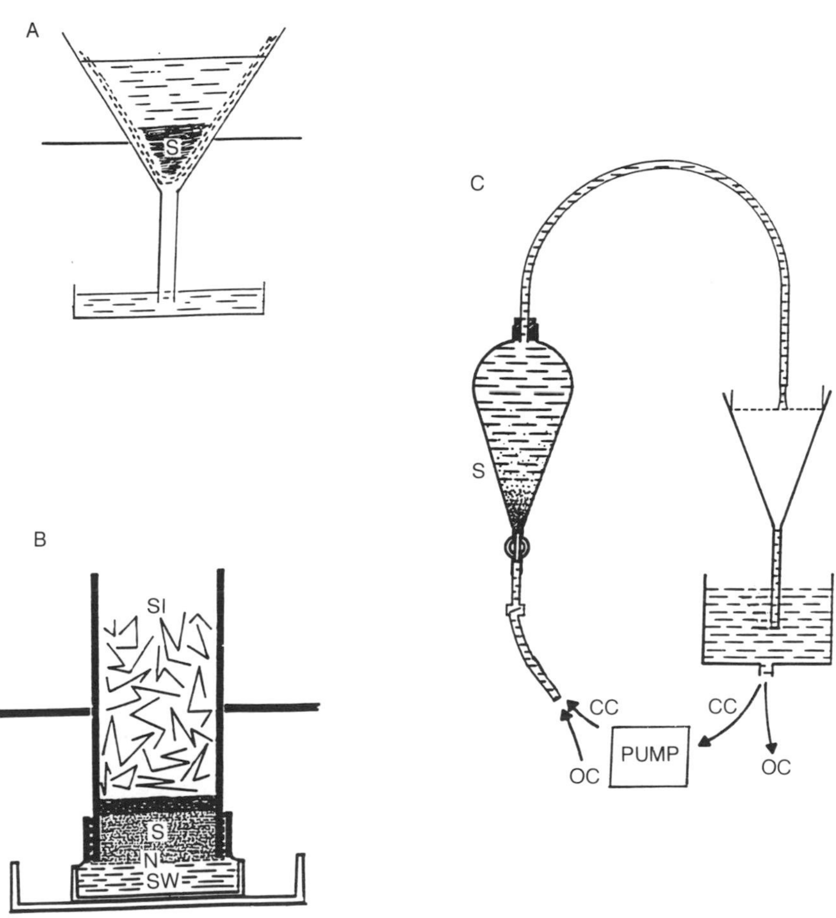

Fig. 16.2. Extraction techniques. (A) Funnel method showing water percolating through sample (S), which is retained on gauze or other wide-mesh net; (B) seawater ice method of Uhlig (1964): seawater ice (SI) is allowed to thaw above sample (S), which is held in place by wide-mesh netting (N); (C) elutriation apparatus showing path of both recirculating or closed cycle (CC) and open cycle (OC) water flow. (B and C modified from Pfannkuche and Thiel, 1988.)

Elutriation

This extraction method relies on water flushed upward through the sample. The sample is place in a closed vessel over a screen and water is forced upward through the sample by a pump. Water passing through is then collected and examined for fauna that have been removed. A more sophisticated version of this method is to recirculate the water for multiple passes through the sample. A catch basin or reservoir acts to concentrate the fauna of interest by sedimentation (Fig. 16.2C).

Flotation

This method has been used with macrofauna and meiofauna in soils and sediments (Giere, 1993). In this approach, the sample is suspended in water with density increased (to a specific gravity of about 1.2 or greater) by addition of sugar or colloidal silica gels ('Ludox AM', 'Ludox TM', 'Ludox LS' or 'Ludox HS' - Du Pont), followed by centrifugation. These methods are designed for samples which have been fixed, although it is possible to use 'Percoll' or 'Percoll-Sorbitol' mixtures, which do not harm the organisms. Again, repeated treatments are often necessary to assure maximal extraction.

Anaesthetics and Fixatives

Anaesthetics and fixatives are often used in conjunction with sampling for sediment fauna. Anaesthetics may improve the efficiency of extraction of those animals with adhesive glands or other mechanisms which secure them tightly to sediment particles, and fixatives will preserve samples for long periods. These may include a variety of compounds (Table 16.2), but, for bdelloid rotifers, which are common in soils, many authors recommend Edmondson's boiling-water method (Edmondson, 1959). Briefly, this involves rapidly adding an equal volume of boiling water to water in which collected rotifers have been placed. Edmondson (1959) notes that not all rotifers in the sample will be well 'relaxed', but that at least some make good specimens. The increase in heat content for relaxation is probably analogous to the increase in temperature used to improve extraction efficiency of the funnel method (Overgaard-Nielsen, 1948). Addition of soda water (5-20% of total volume) or 1% $MgCl_2$ (w/v) have also been used as relaxation techniques. Formalin (a saturated formaldehyde solution) has been the most common fixative, although others have been used (Table 16.2). Pennak (1989) notes that some workers add a small amount of eosin to stain rotifers. Lugol's iodine as a preservative also colours rotifers, and rose bengal may also be used.

The use of either anaesthetics or fixatives must be carefully evaluated in the case of rotifers, however, as there is no universally suitable anaesthetic or

Table 16.2. Selected anaesthetics and fixatives for use with rotifers.

Recommended applications are from Edmondson (1959), Pennak (1989), Wallace and Snell (1991), Nogrady *et al.*(1993).

Compound	Recommended application
Anaesthetics	
Neosynephrin hydrochloride	Add 3-4 drops of 1% (w/v) solution every 5 minutes until ciliary action stops
Novocain	5% (w/v) solution in 50% (v/v) methyl alcohol, applied as neosynephrin
Procaine hydrochloride	0.04% (w/v) concentration for 10-15 hours
Benzamine hydrochloride	Several drops of 2% (w/v) solution added to a small amount of water.
Bupivicaine	Add to achieve final concentration of 0.16 mM
Chloreton	Add to achieve final concentration of 4.0 mM
Hydroxylamine hydrochloride	Add to achieve final concentration of 10.0 mM
Acetone	Add dropwise until activity stops
Magnesium chloride	1% (w/v) final concentration
Soda water (carbonated water)	Add 5 to 20% of total volume
Boiling water	Rapidly add a volume equal to sample volume. Some individuals will relax and fix well
Fixatives	
Formaldehyde	2-5% (v/v) final concentration
Glutaraldehyde	1-2% (v/v) final concentration
Ethanol	30-50% (v/v) final concentration
Lugol's iodine	4-5% (v/v) final concentration

fixative (Nogrady *et al.*, 1993). For many soft-bodied rotifers, those most common in terrestrial systems, fixation (even sometimes after anaesthetization) causes contraction into a ball and renders them useless for taxonomic identification. Turner (1990) circumvented this problem by dividing samples derived from sandy beaches, fixing one subsample directly with formalin, and narcotizing the second subsample with a mixture of one part carbonated water to 20 parts sample water prior to fixation. The first procedure left loricate rotifers in a contracted state, while the second left soft-bodied forms in a relaxed state, which facilitated identification of each type. Thus, it appears that there is still much art in the science of rotifer preservation, and many investigators (e.g. Pourriot, 1979; Wallace and Snell, 1991; Nogrady *et al.*, 1993) recommend observation of live material where possible. If this is not possible, then careful observation of the effect any anaesthetic or preservative used has on taxa commonly found in the sample is essential.

Reporting Results

It is necessary to report results quantitatively. In soils and sediments, results should be presented with respect to the volume of sample and the density of soils and sediments. For hard surfaces, surface area sampled and substrate type should be reported. For plant or detrital material, dry weight of sample should be reported.

Recommended Protocol

Given the difficulties of sampling, extraction, fixation and identification discussed, it is difficult to recommend an ideal protocol for evaluation of sediment or soil rotifers at this time. The following protocol is offered as a suggested starting-point, with the caveat that it must be tested at the specific sampling location to ensure suitability. Further, because evaluation of rotifers is often part of a larger effort that includes many other meio- or mesofaunal components, trade-offs that reduce the efficiency of rotifer sampling and increase the effectiveness of sampling the larger community may be acceptable.

- Collect samples using coring devices. Piston corers are recommended for saturated soils and sediments. The number of replicate cores should be determined by preliminary studies to meet a desired standard, although replicates may be pooled to reduce the number of samples for extraction and counting.
- Transport samples in cooled containers to the laboratory for extraction within 24 h.
- Cores should be sectioned at appropriate levels (e.g. with respect to soil litter layers, soil horizons, anaerobosis) to elucidate vertical structure. Volume and wet weight of each subsample should be recorded.

- Extract cores by elutriation. Volume of water and length of elutriation should be determined empirically by periodically examining aliquots of elutriant to cnsure that all animals have been removed. Use of anaesthetics should also be determined empirically.
- After elutriation, the dry weight of each extracted sediment sample should be noted.
- Fauna from the sample may be concentrated by gentle filtration (20 μm mesh net) or gentle centrifugation of elutriant.
- Relax samples by addition of equal volume of boiling water and fix with formalin.
- Identify and enumerate samples under appropriate magnification (×100-×400) in depression slides. Individual rotifers may be transferred to other slides for treatment with sodium hypochlorite to expose trophi.
- Report results on a volume basis, noting the depth of the layer evaluated and providing sediment wet to dry weight ratios, and other physical and chemical data that may be available.

Limitations of Existing Techniques

Elutriation requires somewhat more sophisticated and expensive equipment (pumps, centrifuges) but may provide the most efficient recovery of fauna, especially in fine-grained sediments. Flotation requires moderately expensive chemicals. Direct sample observation is the most labour-intensive. Funnel filtration may be best in many cases due to simplicity of equipment, set-up and effort, and reports of high extraction efficiency (Pourriot, 1979), at least in coarse-grained sediments. Although rotifers are now recognized as potentially important contributors to most environments, much remains to be learned about their distribution and quantitative importance. Difficulties with taxonomy and appropriate sampling methods remain. Although a protocol is recommended, any investigation should begin with preliminary assessments to determine the most appropriate site-specific sampling design and methods. In general, the optimal method for extraction of samples in a given study will vary with the nature of the site and the resources available.

Acknowledgements

I am grateful to R.L. Wallace, who provided constructive criticism which improved the manuscript.

Bibliography

General Texts

Edmondson, W.T. (1959) *Rotifera*. In: Edmondson, W.T. (ed.) *Fresh-water Biology*, 2nd edn. Wiley, New York, pp. 420-494. [General text with taxonomic key and extensive reference list.]

Giere, O. (1993) *Meiobenthology*. Springer-Verlag, Berlin. [General text.]

Nogrady, T., Wallace, R.L. and Snell, T.W. (1993) *Rotifera*. Vol. 1: *Biology, Ecology and Systematics*. SPB Academic Publishing, The Hague. [Excellent general discussion of rotifer biology and ecology with extensive up-to-date reference list.]

Pennak, R.W. (1989) *Freshwater Invertebrates of the United States*. John Wiley & Sons, New York. [General text with rotifer taxonomic key.]

Pourriot, R. (1979) Rotifères du sol. *Revue d'Ecologie et de Biologie du Sol* 16, 279-312. [Most complete review of soil rotifers.]

Wallace, R.L. and Snell, T.W. (1991) *Rotifera*. In: Thorp, J. and Covich, A. (eds) *Ecology and Classification of North American Freshwater Invertebrates*. Academic Press, New York, pp. 187-248. [General review of rotifer biology with taxonomic key.]

Techniques

Fleeger, J.W., Thistle, D. and Thiel, H. (1988) Sampling equipment. In: Higgins, R.P. and Thiel, H. (eds) *Introduction to the Study of Meiofauna*. Smithsonian Institution Press, Washington DC, pp. 115-125.

May, L. (1986) Rotifer sampling - a complete species list from one visit? *Hydrobiologia* 134, 117-120.

Myers, F.J. (1937) A method of mounting rotifer jaws for study. *Transactions of the American Microscopical Society* 56, 256-257.

Overgaard-Nielsen, C. (1948) An apparatus for quantitative extraction of nematodes and rotifers from soil and moss. *Natura Jutland* 1, 271-277.

Peters, U., Koste, W. and Westheide, W. (1993) A quantitative method to extract moss-dwelling rotifers. *Hydrobiologia* 255/256, 339-341.

Pfannkuche, O. and Thiel, H. (1988) Sample processing. In: Higgins, R.P. and Thiel, H. (eds) *Introduction to the Study of Meiofauna*. Smithsonian Institution Press, Washington DC, pp. 134-145.

Uhlig, G. (1964) Eine einfache Mëthode zur Extraktion der vagilen mesopsammen Mikrofauna. *Helgolander Wissenschaftliche Meeresuntersuchungen* 11, 178-195.

Specific Texts

Anderson, R.V., Ingham, R.E., Trofymow, J.A. and Coleman, D.C. (1984) Soil mesofaunal distribution in relation to habitat types in shortgrass prarie. *Pedobiologia* 26, 257-261.

Evans, W.A. (1982) Abundances of micrometazoans in three sandy beaches in the Island area of western Lake Erie. *Ohio Journal of Science* 82, 246-261.

Evans, W.A. (1984) Seasonal abundance of the psammic rotifers of a physically controlled stream. *Hydrobiologia* 108, 105-114.

Oden, B.J. (1979) The freshwater littoral meiofauna in a South Carolina reservoir receiving thermal effluents. *Freshwater Biology* 9, 291-304.

Palmer, M.A. (1990) Temporal and spatial dynamics of meiofauna within the hyporheic zone of Goos Creek, Virginia. *Journal of the North American Benthological Society* 9, 17-25.

Pennak, R. (1988) Ecology of the freshwater meiofauna In: Higgins, R.P. and Thiel, H. (eds) *Introduction to the Study of Meiofauna*. Smithsonian Institution Press, Washington DC, pp. 39-60.

Rublee, P.A. and Partusch-Talley, A. (1995) Microfaunal response to fertilization of an arctic tundra stream. *Freshwater Biology* 34 (in press).

Schmid-Araya, J.M. (1993) The benthic *Rotifera* inhabiting the bed sediments of a gravel brook. *Jahresbericht Biologisches Station Lunz* 14, 75-101.

Sohlenius, B. (1979) A carbon budget for nematodes, rotifers and tardigrades in a Swedish coniferous forest soil. *Holarctic Ecology* 2, 30-40.

Sohlenius, B. (1982) Short-term influence of clear-cutting on abundance of soil microfauna (*Nematoda, Rotatoria* and *Tardigrada*) in a Swedish pine forest soil. *Journal of Applied Ecology* 19, 349-359.

Thane-Fenchel, A. (1968) Distribution and ecology of non-planktonic brackish-water rotifers from Scandinavian waters. *Ophelia* 5, 273-297.

Turner, P.N. (1990) Some interstitial rotifera from a beach in Florida, USA. *Transactions of the American Microscopical Society* 109, 417-421.

Tzschaschel, G. (1980) Verteilung, Abundanzdynamik und Biologie mariner interstitieller Rotatoria. *Mikrofauna Meeresboden* 81, 1-56.

Turbellarians 17

Ernest R. Schockaert

Limburgs Universitair Centrum, Zoology Research Group, Departement SBG, B-3590 Diepenbeek, Belgium.

The Importance of Turbellarians in Ecosystems

Though the turbellarians are considered a class in most handbooks, they have been recognized as a paraphyletic group and have therefore lost the status of a taxon. Nevertheless, the term remains useful to indicate the ciliated *Platyhelminthes*, of which the majority are free-living. Approximately 5500 species are known, which are distributed over 10-11 orders (Ehlers, 1985; Fig. 17.1), mostly ranked as ordines in traditional handbooks.

Turbellarians are known from freshwater and from marine environments, while *Terricola* (*Tricladida*) are exclusively terrestrial in damp environments, although many so-called 'microturbellarians' may also be found there. Larger species, such as most of the *Polycladida* and *Tricladida*, are epibenthic (mostly on or under stones and rocks); many species are found on or around water, submerged plants and algae or live in sand.

Very few freshwater turbellarians occur in sediments, and attention will mainly be paid to the marine sand-inhabiting species, although some comments are given on freshwater turbellarians as well. Most of the techniques discussed are valid for the study of freshwater turbellarians, or can be adapted.

Most sand-dwelling turbellarians are interstitial, i.e. live in the spaces between the sand grains. However, many of them are rather large, and displace the sand granules. Whether they are to be considered meiobenthic depends on the definition of meiobenthos which is adopted. If the size criterion in used (± 40 µm to 1 mm *fide* Vincx, this volume, Chapter 15), then some representatives of some taxa are meiobenthic, but others are not. If life strategy is the criterion, such as life cycle being completely in the sand (Warwick, 1984), then turbellarians are definitely meiobenthic, except for the *Polycladida*, some of which have a larval stage (but which are seldom found in sand anyway). As this problem is also present in a number of other taxa such as *Nematoda*, *Gastrotricha*, some '*Archiannelida*', the life strategy definition has been adopted here.

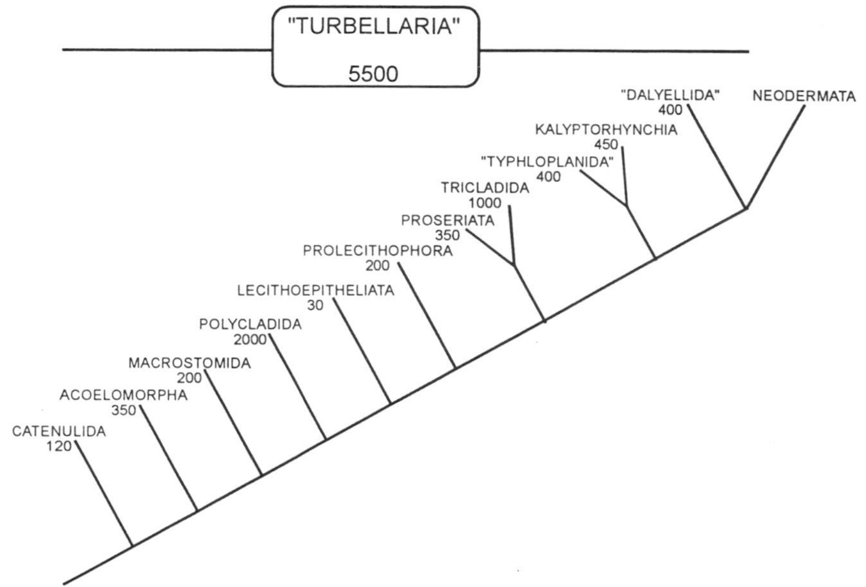

Fig. 17.1. Cladogram of the *Platyhelminthes* with the approximate numbers of known species per turbellarian order.

Densities of turbellarians in marine sediments may be surprisingly high and numbers of 1600 individuals 10 cm^{-2} have been recorded in sand in a mangrove area (Alongi, 1987). In Belgian sandy beaches over a period of 15 months, peak densities of 380 individuals 10 cm^{-2} with a relative abundance of 37% of the meiofauna were recorded (Jouk, 1992). Quantitative data from the Belgian coast and from the Island of Sylt (Jouk, 1992; Armonies and Hellwig-Armonies, 1987) show that 20% or more of turbellarians in the total number of the meiobenthic animals is not at all exceptional. Instability of the habitat (exposure for beaches) apparently causes a decrease of the total meiobenthos, but a relative increase of turbellarians (Martens *et al.*, 1985). The density of turbellarians in sublittoral habitats is mostly much lower than in intertidal localities.

Nematodes are undoubtedly the most abundant taxon in the meiobenthos and turbellarians are often the second most abundant taxon, more often third after harpacticoids or sometimes gastrotrichs. Data on the biomass of turbellarians are very scarce. Faubel (1982) weighed a number of turbellarians and related the dry weight to the size of the animals. Jouk (1992) also weighed all major taxa found in the sand of the Belgian beaches and related weight to size. The range of the individual dry weight for turbellarians is 0.62-11.0 μg dry weight per individual. Dry weight of some taxa, and even the weight of individual species, may be slightly different from locality to locality (but size varies similarly). In two Belgian beaches, the total biomass represented by the turbellarians was found to be higher than that of nematodes.

Quantitative estimates of the diversity of turbellarians are only known from habitats in the neighbourhood of the Island of Sylt and from Belgian sandy beaches (e.g. Reise, 1984; Wehrenberg and Reise, 1985; Jouk, 1992). Expressed as a Shannon-Wiener index (for example), values over three were recorded, the average being mostly above two, seldom less than one, in intertidal as well as in sublittoral habitats. The number of species may be about 30 in one single sample and in one locality on the Belgian coast 70 species were found throughout the year.

The ecological importance and impact of turbellarians in the marine meiofauna have been discussed by Martens and Schockaert (1986) and extensive data on the ecology of marine turbellarians can be found in Reise (1984, 1987), Armonies and Hellwig-Armonies (1987), Armonies (1987) and Jouk (1992)

Most turbellarians are generally confined to the aerobic zone of the sediment and, in littoral stations, greatest densities are found in the upper 2-5 cm (even when the anoxic layer is much deeper), although we found large numbers of turbellarians in the black layer (from 2 cm down) in sand in a tropical mangrove area. Vertical migrations in the sediment by the meiofauna have been noted (see Xylander and Reise, 1984, for references), but have not been observed in German or Belgian localities studied so far. In sublittoral sandy habitats, they may occur deeper in the sediment. Although turbellarians may be considered interstitial, many species also swim. As a rule, the finer the sediment (or the lower the dynamics in the habitat), the more swimming species occur. In sand, the majority of species strongly adhere to the grains. In contrast, freshwater turbellarians do not display adhesive behaviour, apart from a few taxa.

Most marine turbellarians have a univoltine life cycle (i.e. one reproductive period per year), few are bivoltine or polyvoltine (Ax, 1977; Hellwig, 1987). The life cycles of all species are not synchronized and the densities of the whole turbellarian fauna and the relative abundance of the various species may vary considerably over time (Armonies, 1990; Jouk, 1992). In brackish water, some species react to the unstable environment by the formation of cysts, while in fresh water many species produce subitaneous eggs (i.e. ones which develop immediately) in summer (or in the wet season), and resting eggs when the unfavourable season approaches (see e.g. Heitkamp, 1988). Relatively recent studies on the biology and the ecology of freshwater turbellarians, including some sediment habitats, are given in Schwank (1981a,b; 1982a,b; 1986), Kolasa *et al.* (1987) and Kolasa (1991). Schwank's list of truly interstitial freshwater turbellarians in *Stygofauna Mundi* (1986) and the checklist of European freshwater turbellarians by Lanfranchi and Papi (1978) may be useful.

From the small selection of data presented above it must be sufficiently clear that turbellarians are worthy of consideration in biodiversity and ecological studies. In marine sediments, diversity and biomass are in the same range as those of nematodes. As surface predators, they may also play a major role in other aspects in the ecology of permanent and temporary meiobenthic communities (Martens and Schockaert, 1986).

Methods for Sampling/Extraction of Turbellarians

As for most soft-bodied animals, turbellarians should be studied alive for proper identification. Turbellarians extracted from fixed sediment samples (i.e. for quantitative studies) can be identified only if a fairly good knowledge of the species in the habitat under investigation is available. A preliminary study of qualitative samples from a new habitat is highly advised, since even in the most intensively studied European North Sea sediments (littoral and sublittoral) unknown species can be expected.

Qualitative Sampling

For a qualitative study of the live turbellarian fauna of sandy littoral stations, a surface area is scraped of to a depth of about 5-10 cm and transported in a bucket or plastic bag. In muddy or freshwater habitats, only a thin upper layer is collected and preferably transported in a glass jar with lid.

Sublittoral samples can be taken with any sampling device such as grabs, various corers or even dragging nets (sledge), provided sediment is brought in from which subsamples can be taken. Various devices are described in Fleeger *et al.* (1988). Since turbellarian densities in sublittoral habitats may be expected to be lower than in the intertidal, relatively large samples should be used.

Quantitative Sampling

For quantitative sampling, methods and strategies generally used for the study of the marine meiobenthos can be used (Fleeger *et al.,* 1988; Vincx, this volume, Chapter 15). The German group (Reise, Armonies, Hellwig, Xylander and others) uses 2 cm^2 cores directly taken or 1 cm^2 subsamples from a 10 cm^2 core and counts 10 or 20 replicates of these small cores.

In the opinion of the author, sampling method and strategy for quantitative studies of the turbellarians should follow general procedures used for meiobenthos (see Vincx, this volume, Chapter 15) as closely as possible to minimize distortions in comparisons. Small cores, however, have been proved very useful in studies on niche partitioning where only turbellarians were concerned (Reise, 1984, 1987).

Qualitative Methods for the Extraction of Living Turbellarians

The decantation method

A description of extraction methods for living specimens can be found in Pfankuche and Thiel (1988) and, in our experience, the decantation method with $MgCl_2$ is the most practical one to use routinely (Fig. 17.2).

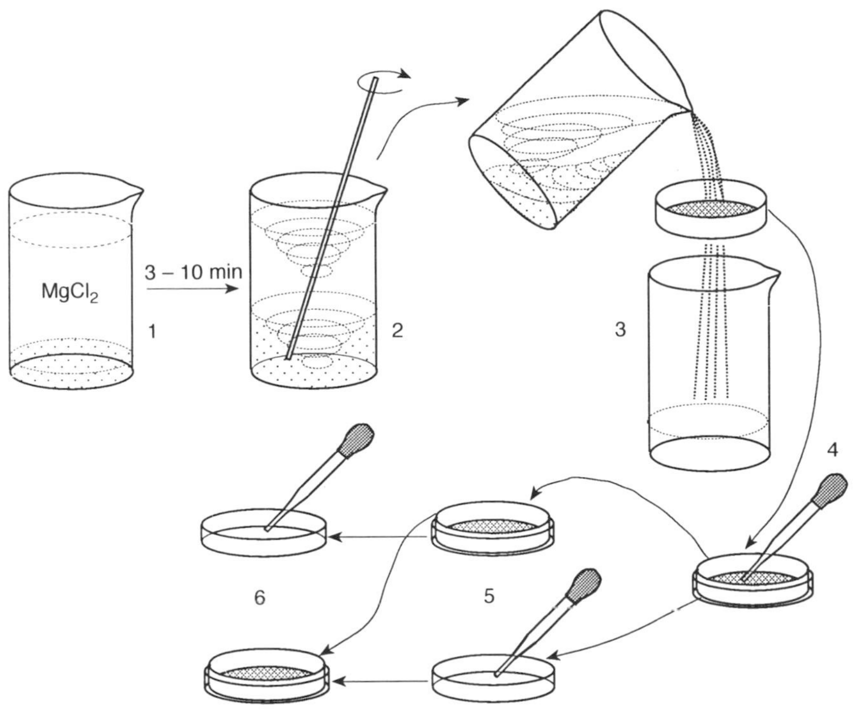

Fig. 17.2. The decantation method. Numbers on this figure refer to those given in the procedure described.

1. Put a 3 cm layer of sediment in a tall, narrow 1.5 litre capacity beaker (glass or plastic) and 1 litre of isotonic $MgCl_2$ solution is added (*ca.* 75 g $MgCl_26{:}H_20$ in 1 litre of tap water).
2. Wait 3-10 min for anaesthetization and stir (not too vigorously so as not to damage fragile species).
3. Decant the supernatant (with the floating animals and as little sediment as possible) over a 65-95 µm mesh plankton net glued on a plastic ring that fits into a Petri dish.

4. Place the net in a Petri dish and add seawater (filtered if available). Wait a minute or so until the animals recover from the anaesthetic. Most turbellarians (like many other mobile meiofauna elements) seem to be negatively geotropic and leave the net trough the meshes.
5. After some time, remove the net and transfer it to a second Petri dish with seawater. Many individuals can now be found in the first Petri dish, almost completely free of sand.
6. When all the species in this first Petri dish have been examined, place the net back into this dish and pick out the animals in the second dish, and so on until all the species have been recovered.

Not all species move equally fast through the net; some are very slow, or prefer to swim, and it is advisable to control the net carefully. The sediment that has been extracted can be extracted several times. The sediment can be left in the $MgCl_2$ while you are sorting the first extraction. The same $MgCl_2$ solution can be used several times until it is too contaminated with detritus or the finer sediment fraction. In order to save $MgCl_2$, it can be useful to perform the first extraction(s) with seawater alone (in which some species can be extracted). In very fine sediment, several subsequent extractions can be done with seawater alone and $MgCl_2$ used at the very last stage since less adhesive species occur in this kind of sediment. Some authors advise the use of an ethanol solution (up to 5-10% v/v), or a weak formalin solution (instead of the $MgCl_2$ solution), which may also be used in freshwater samples. Fresh water can also be used for 10-15 s to extract marine samples, but it may damage some species and seems less effective than $MgCl_2$.

The seawater ice method

Another handy method to extract living turbellarians from the sediment is the seawater ice method of Ulhig *et at.* (1973) (see also the description in Pfankuche and Thiel (1988) and in Rublee, this volume, Chapter 16). It is slightly more effective than the $MgCl_2$ method and has the advantage of bringing less sediment into the Petri dish, but is more time-consuming. A net of 100 μm mesh or less should be used.

The oxygen depletion method

For the extraction of living turbellarians from very fine and silty sediments, the oxygen depletion method may also be useful, although it is not equally effective for all taxa. A layer of sediment in a high glass beaker is covered with seawater and left at room temperature without being moved. Turbellarians tend to crawl upward on the glass wall or to swim and can be picked up with a pipette. Any disturbance of the beaker will cause them to fall down. This method has proved to be useful in brackish water habitats such as salt marshes and is the appropriate method for freshwater turbellarians. The surface layer of the sediment can also be aspirated and viewed for turbellarians under the stereomicroscope.

The oxygen depletion method can be used in combination with the $MgCl_2$ or seawater ice method. For sandy sediments, the bucket with the qualitative sample may be left at room temperature for a few days, when the animals concentrate in the upper layer, which is than treated for extraction. Armonies and Hellwig (1986) described a simple method to extract several meiofaunal elements, among which were turbellarians, from muddy sediments. A small mud sample, several centimetres thick, is covered with clean, dry, coarser sand that has been moistened with seawater. When kept at room temperature, preferably in the dark, the turbellarians tend to move upward in the coarse sand, from which they can then be extracted. This treatment should be repeated several times, and according to the authors, it works quantitatively as well. In very muddy sediments, sieving the sediment under running seawater over a 45-65 μm net to get rid of the finest fractions may prove a workable method.

Flotation-decantation methods

'Percoll'-sorbitol or sucrose (680 g l^{-1}) combined with centrifugation have been described to be effective for live meiofauna (Pfankuche and Thiel, 1988). Other methods are described by Rublee (this volume, Chapter 16). I have no experience with those methods and to the best of my knowledge they have never been used to extract turbellarians.

Qualitative samples from littoral or shallow water habitats from temperate, tropical and subpolar areas may be kept for several days at room temperature or in a cool place (10-20°C) before examination, although the *relative* abundance of the various species present may change. Samples from deeper localities are more sensitive and should be treated at low temperature, and it may be necessary to keep the Petri dish on ice while picking out the animals. To study those animals under the microscope may be problematic since they disintegrate very fast.

Quantitative Extraction Methods

It has been claimed that the seawater ice method is a quantitative extraction method for living turbellarians. However, it has been shown by Martens (1984) that this may be true only for the very fragile taxa, such as *Acoela* and *Retronectida*. For all other turbellarian taxa, the extraction of fixed samples yields far higher numbers. The German group extracts the turbellarians alive from the quantitative samples (small cores), then counts and identifies them alive. The shaking-decantation method is used (up to 10 times, ending with $MgCl_2$ or fresh water and control of the remaining sediment). In a muddy sediment, the method of Armonies and Hellwig (1986) described above can be used. However, this is only possible when nearby sites have to be examined, as the method is extremely time-consuming and extraction may be difficult in a dynamic environment with a high number of very adhesive species. Extraction from fixed samples is mostly unavoidable for practical reasons and is probably more effective (in spite of the identification problems which may arise) since a greater amount of sediment (10 cm^2 cores) can be treated.

Various extraction methods for fixed samples are known (Pfankuche and Thiel, 1988), but, in our experience, the method described by Barnett (1986) produces very good results, as does the shaking-decantation method - 10 times followed by three extractions with a flotation-centrifugation on 'Ludox' or sugar (Vincx, this volume, Chapter 15). **N.B.** Flotation-centrifugation with 'Ludox' seems to damage turbellarians to some degree.

Sorting

For sorting the extracted live animals from the Petri dishes, a stereomicroscope with minimum magnification of about ×2.5 should be used. A cold light source with two flexible glass fibre arms is recommended to avoid overheating the animals. Otherwise, adapting the distance of the light source, or using a heat filter, and certainly switching off the light when not observing should be considered. Whether incident or transmitted light is used is a matter of taste, but the author has found a combination of both to be most useful, e.g. a stereomicroscope on a stand with a mirror that allows both techniques to be employed. With incident light on a black background the animals appear whitish and are easily found between the sand grains. Be aware of the fact that representatives of some taxa (such as many *Kalyptorhynchia*) often remain quietly attached to the bottom of the Petri dish.

Specimens can be picked up with a disposable Pasteur pipette with a rubber bulb. Some workers use a pipette with a tube and a mouthpiece, or an 'Erwin-loop' (a small metal loop on a holder; Westheide and Purschke, 1988); **never use a needle.** Sand-dwelling turbellarians are often very sticky and are not always easy to remove from the bottom of the Petri dish (or from a sand grain), and it is best to wait until the animal is moving about, which may require some patience. **Do not disturb it:** the more the animal is disturbed, the more it sticks! Sticking to the pipette may be a problem as well, and, to avoid this, the inside of the pipette may be coated with silicone, but, in my experience, very fast transfer of the specimen mostly works well.

Animals are best transferred to a watch-glass, a cavity block (embryo dish) or a finger bowl, where they can be concentrated. Remove surplus of water, sand and dirt from time to time. Immediate transfer to a slide for microscopic study is possible provided not too much water is transferred as well. Sticky species can be picked up more easily from the cavity block, since they can be removed from the bottom by blowing them off without losing them from the field of view of the stereomicroscope. Before transferring the animal to the slide some notes (or even drawings) on its behaviour should be made as well as the colour under incident light.

Mounting

More or less sophisticated (and expensive) devices have been designed to study living animals under the compound microscope (Westheide and Purschke, 1988) but, in the author's experience, a normal flat slide and a small cover glass (18 × 18 mm, or small circular) are sufficient. A larger cover glass may be too heavy and make most species burst. For very fragile or for very thick species, 'Plasticine' or wax 'feet' at the edges may sometimes be useful. (Scrape off the 'Plasticine' directly with the edges of the cover glass; the four feet should be equally thick since image distortion may occur otherwise).

The specimen is transferred from the cavity block (or the Petri dish) onto a slide (Fig. 17.3) and surplus water removed, leaving sufficient so that the water remains completely under the coverslip and cannot escape from underneath, but the animal is not squashed either (step 1). This may require some prior training and the more water should remain the larger the animal is. Be particularly careful to remove all sand grains or any dirt, since these may prevent adequate compression of the animal later on, and so it is better to use the stereomicroscope at low magnification.

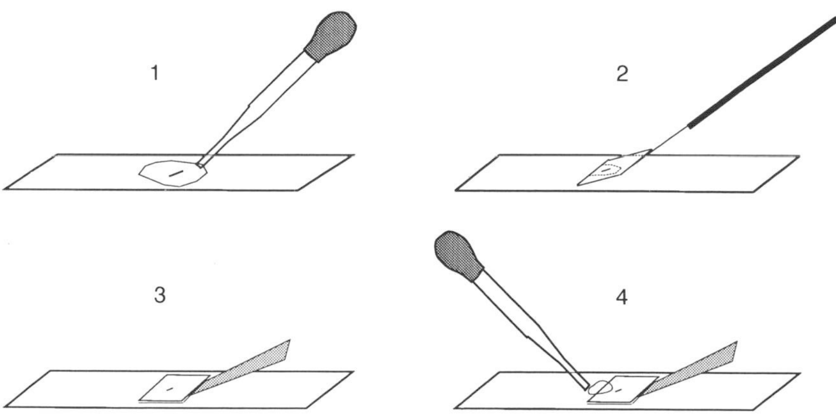

Fig. 17.3. Preparation of the mount. Numbers refer to steps described in the text.

The water droplet with the animal is now covered with the cover slip (step 2), which is set on one side on the slide and allowed to glide down very gently using fine forceps or a needle, without damaging the animal and avoiding air bubbles. More water can be removed **very gradually** from underneath the cover slip with filter paper (step 3) until the animal slows down but it should **not** be immobilized. Some anaesthetic (e.g. $MgCl_2$ solution) may be sucked underneath the cover slip as well, but after some experience this will appear unnecessary.

Observations and Records for Identification

To study the living turbellarians a good-quality compound microscope is needed. A ×3-4 objective (in order to find the moving animal easily), a ×100 immersion objective and intermediate magnifications of ×10, ×20 and ×40 or ×60 are highly recommended. Bright-field microscopy may be used, and phase-contrast, interference-phase-contrast or Nomarski DIC optics may be useful, but are not required.

If the species is unknown, all necessary data should be recorded for later identification. The animal is first measured at low magnification, preferably before it is under the cover glass, but this measurement is only rather approximate, with a precision of about 0.5 mm. Note the colour (if any) under the microscope: it is mostly different from that under the stereomicroscope. While the animal is still moving, important data can be recorded at low magnification (×10 or ×20) such as: the position of the pharynx, the presence and type of proboscis, the position of the testes, ovaries, vitellaria and genital pore. Many of these structures will become less clear during the squeezing process or may change their positions or proportions. On the other hand, many structures are not identifiable at this point, but will become obvious later. A drawing of the whole animal as accurately as possible should be made and a photograph is very useful at this stage (high-speed emulsion of 400 ASA).

While studying the specimen, the preparation dries slowly and the animal is more and more squeezed and becomes less and less mobile. More and more internal structures become visible and may be viewed at higher magnification. Squeezing can be speeded up by removing more water with filter paper, but this should be done very gradually. When the squeezing process is carried out too fast, many structures become (rather suddenly) unrecognizable and poorly delimited. A drawing (and photograph) of the reproductive system at a higher magnification (×40 or ×60 objective) should be made. Do not pay too much attention to details of the hard parts in the copulatory organ (if there are any), as they can be studied in a permanent mount, although the general shape should be recorded. These last details can be observed with the immersion objective and added to the drawing.

Whether all data are recorded as drawings or as a written report depends or your personal taste and talents. Photography (or video recording) is useful, but cannot replace direct observation.

Permanent Mounts

Once all necessary data have been recorded on the living animal, which will probably be dead by now or perhaps even completely disintegrated, a permanent mount for the detailed study of the hard parts can be made. A droplet of lactophenol[1] is put at one side of the coverslip and sucked underneath using filter paper from the other side (step 4 in Fig. 17.3). The approaching liquid front can be monitored under the microscope or stereo microscope. Once the lactophenol is completely under the cover slip, all surplus lactophenol is removed with the filter paper (without introducing air bubbles). For freshwater animals, polyvinyl lactophenol can be used, which hardens gradually, but does not dissolve in seawater and cause precipitation. Mounts in lactophenol may be sealed with polyvinyl lactophenol applied with a brush around the cover slip, or with resin - 'Glyceel' does not work. Sealing is not absolutely necessary, but it makes the preparation less vulnerable to distortion or damage, as the cover slip cannot then be moved.

These whole mounts can now be studied at any time and fine details of the hard parts recorded. In individuals which have remained intact, some of the soft parts can still be seen with Nomarski DIC optics, but these data alone are seldom sufficient for identification.

Mounts of fixed specimens (e.g. extracted from quantitative samples) can be made in the same way. After some days in lactophenol (under the cover glass), the animal is completely cleared. Thick species may require some pressure (e.g. under a small metal block) and leaving them in a 15% (v/v) acetic acid solution for some hours to soften the tissues has proven very useful. (Monitoring the specimen is necessary to ensure that it does not disintegrate.)

The majority of the species can be identified by the hard parts in the reproductive system. Most taxa without hard parts can be identified as well (with Nomarski DIC optics), although, if many species of the same higher taxon (e.g. acoels) are present, this may become tedious. It should be understood that identification of these fixed individuals is more a process of allocating them to the species which have previously been identified alive!

Identification

The pictorial key (to family and genus level in some families) of Cannon (1986) may prove very useful in the first instance. Monographs on various taxa can be used as well and some are listed separately in the Bibliography here. For some taxa without hard parts in the reproductive system (e.g. a number of *Acoela*, all *Prolecithophora*) histological sections and reconstruction may be required for identification. (Not many prolecithophoran species occur in sediments, however, but one or two may be found in samples.) Sectioning and reconstructing is mostly required for a good description of a new genus (even species sometimes). However, descriptions of these techniques are beyond the scope of this work, although some can be found in Westheide and Purschke (1988).

---

[1]Editor's Footnote: CAUTION! Lactophenol is a potential, although unproven, carcinogen. It may be possible to substitute lactic acid (?% v/v) for lactophenol.

Acknowledgements

I would like to thank Dr J. Kolasa (McMaster University, Ontario, Canada) for his advice and suggestions for the study of freshwater turbellarians.

Bibliography

Alongi, D.M. (1987) Intertidal zonation and seasonality of meiobenthos in tropical mangrove estuaries. *Marine Biology* 96, 447-458.

Armonies, W. (1987) Freilebende Plathelminthes in supralitoralen Salzwiezen der Nordsee: ökologie einer borealen Brackwasser-lebensgemeinschaft. *Microfauna Marina* 3, 81-156.

Armonies, W. (1990) Short-term changes of meiofaunal abundance in intertidal sediments. *Helgoländer Meeresuntersuchungen* 44, 375-386.

Armonies, W. and Hellwig, M. (1986) Quantitative extraction of living melofauna from marine and brackish muddy sediments. *Marine Ecology Progress Series* 29, 37-43.

Armonies, W. and Hellwig-Armonies, M. (1987) Synoptic patterns of meiofaunal and macrofaunal abundances and specific composition in littoral sediments. *Helgoländer Meeresuntersuchungen* 41, 83-111.

Ax, P. (1977) Life cycles of interstitial *Turbellaria* from the eulitoral of the North Sea. *Acta Zoologica Fennica* 154, 11-20.

Barnett, P.R.O. (1986). Distribution and ecology of harpactoid copepods of an intertidal mudflat. *Internationale Revue Gesamten Hydrobiologie* 53, 177-209.

Cannon, L.R.G. (1986) *Turbellaria of the World: A Guide to Families and Genera*. Queensland Museum, Brisbane.

Ehlers, U. (1985) *Das phylogenetische System der Plathelminthes*. Fischer, Stuttgart.

Faubel, A. (1982) Determination of individual meiofauna dry weight values in relation to definite size classes. *Cahiers de Biologie Marine* 23, 339-345.

Fleeger, J.W., Thistle, D. and Thiel, H. (1988) Sampling equipment. In: Higgins, R.P. and Thiel, H. (eds) *Introduction to the Study of Meiofauna*. Smithsonian Institution Press, Washington DC, pp. 115-125.

Heitkamp, U. (1988) Life-cycles of microturbellarians of pools and their strategies of adaptation to their habitats. *Progress in Zoology* 36, 449-456.

Hellwig, M. (1987) Ökologie freilebenden Plathelminthen im Grenzraum Watt-Salzwiese lenitischer Gezeitenkusten. *Microfauna Marina* 3, 157-248.

Jouk, P. (1992) *Study of the ecology of free-living Platyhelminthes ('Turbellaria') in sandy beaches of the Belgian coast*. PhD thesis, Limburg University Centre, Diepenbeek, Belgium.

Kolasa, J. (1991) *Flatworms: Turbellaria and Nemertea. Ecology and Classification of North American Freshwater Invertebrates.* Academic Press, New York, pp. 145-171.

Kolasa, J., Strayer, D. and Bannon-O'Donnel, E. (1987) Microturbellarians from interstitial waters, streams, and springs in southeastern New York. *Journal of the North American Benthological Society* 6, 125-132.

Lanfranchi, A. and Papi, F. (1978) *Turbellaria* (excl. *Tricladida*). In: Illies, J. (ed.) *Limnofauna Europaea*. Fisher, Stuttgart, pp. 5-15.

Martens, P.M. (1984) Comparision of three different extraction methods for *Turbellaria*. *Marine Ecology Progress Series* 14, 299-234.

Martens, P.M. and Schockaert, E.R. (1986) The importance of turbellarians in the marine meiobenthos: a review. *Hydrobiologia* 132, 295-303.

Martens, P.M., Jouk, P., Huys, R. and Herman, R. (1985) Short note on the relative abundance of the *Turbellaria* in the Meiofauna of sandy habitats in the Southern Bight of the North Sea and on Belgian Beaches. In: Grieken, R. van and Wollast, R. (eds) *Progress in Belgian Oceanographic Research.* Royal Academy of Sciences, Brussels, pp. 341-342.

Pfankuche, O. and Thiel, H. (1988) Sample processing. In: Higgins, R.P. and Thiel, H. (eds) *Introduction to the Study of Meiofauna.* Smithsonian Institution Press, Washington DC, pp. 134-145.

Reise, K. (1984) Free-living *Platyhelminthes* (*Turbellaria*) of a marine sand flat: an ecological study. *Microfauna Marina* 1, 1-62.

Reise, K. (1987) Spatial niches and long-term performance in meiobenthic Platyhelminthes of an intertidal lugworm flat. *Marine Ecology Progress Series* 38, 1-11.

Reise, K. (1988) Platyhelminth diversity in littoral sediments around the island of Sylt in the North Sea. *Progress in Zoology* 36, 469-480.

Schwank, P. (1981a) Turbellariën, Oligochaeten und Archianneliden der Breitenbachs und anderer oberhessischer Mittelgebirgsbäche. I. Lokalgeographische Verbreitung und die Verteilung der Arten in den einzelnen Gewässern in Abhängigkeit vom Substrat. *Archiv fiir Hydrobiologie, Supplement* 62, 1-85.

Schwank, P. (1981b) Turbellariën, Oligochaeten und Archianneliden der Breitenbachs und anderer oberhessischer Mittelgebirgsbäche. II. Die Systematik und Autökologie der einzelnen Arten. *Archiv für Hydrobiologie, Supplement* 62, 86-147.

Schwank, P. (1982a) Turbellariën, Oligochaeten und Archianneliden der Breitenbachs und anderer oberhessischer Mittelgebirgsbäche. III. Die Taxozönoser der Turbellariën und Oligochaeten in Fliessgewässer - eine synökologische Gliederung. *Archiv für Hydrobiologie, Supplement* 62, 191-253.

Schwank, P. (1982b) Turbellariën, Oligochaeten und Archianneliden der Breitenbachs und anderer oberhessischer Mittelgebirgsbäche. IV. Allgemeine

Grundlagen der Verbreitung von Turbellariën unde Oligochaeten in Fliesgewässern. *Archiv für Hydrobiologie, Supplement* 62, 254-290.

Schwank, P. (1986) Microturbellaria from subterranean freshwater habitats. In: Botosaneanu, L. (ed.) *Stygofauna Mundi*. E.J. Brill, Leiden, pp. 47-56.

Uhlig, G., Thiel, H. and Gray, J.S. (1973) The quantitative separation of meiofauna. A comparison of methods. *Helgoländer Wissenschaftliche Meersuntersuchungen* 25, 173-195.

Warwick, R.M. (1984) Species size distribution in marine benthic communities. *Oecologia* 61, 32-41.

Wehrenberg, C. and Reise, K. (1985) Artenspektrum und Abundanz freilebender Plathelminthes in sublitoralen Sanden der Nordsee bei Sylt. *Microfauna Marina* 2, 163-180.

Westheide, W. and Purschke, G. (1988) Organism processing. In: Higgins, R.P. and Thiel, H. (eds) *Introduction to the Study of Meiofauna*. Smithsonian Institution Press, Washington DC, pp. 146-160.

Xylander, W.E.R. and Reise, K. (1984) Free-living Platyhelminthes (Turbellaria) of a rippled sand bar and a sheltered beach: A quantitative comparison at the Island of Sylt (North Sea). *Microfauna Marina* 1, 257-277.

Monographs (Often with Identification Keys and Checklists)

Curini-Galletti, M.C. and Martens, P.M. (1991). Systematics of the *Unguiphora* (*Platyhelminthes: Proseriata*). I. Genus *Polystyliphora* Ax, 1958. *Journal of Natural History* 25, 1089-1100.

Curini-Galletti M.C. and Martens, P.M. (1992) Systematics of the *Unguiphora* (*Platyhelminthes: Proseriata*). II. *Nematoplanidae* Meixner 1938. *Journal of Natural History* 26, 285-302.

Karling, T.G. (1974) Turbellarian fauna of the Baltic proper: identification, ecology and biogeography. *Fauna Fennica* 27, 1-101.

Luther, A. (1950) Die Dalyeliiden. *Acta Zoologica Fennica* 87.

Martens, P.M. and Curini-Galletti, M.C. (1989) *Monocelididae* and *Archimonocelididae* (*Platyhelminthes: Proseriata*) from South Sulawezi (Indonesia) and Northern Australia with biogeographical remarks. *Tropical Zoology* 2, 175-205.

Martens, P.M. and Curini-Galletti, M.C. (1993) Taxonomy and phylogeny of the *Archimonocelidae* Meixner, 1938 (*Platyhelminthes: Proseriata*). *Bijdragen tot de Dierkunde* 63, 65-102.

Meixner, P. (1938) *Platyhelminthes: Proseriata*. *Bijdragen tot de Dierkunde* 63, 65-102.

Noldt, U. (1989) *Kalyptorhynchia* (*Platyhelminthes*) from sublittoral coastal areas near the Island of Sylt (North Sea.) I. *Schizorhynchia*. *Microfauna Marina* 5, 7-85.

Noldt, U. (1989) *Kalyptorhynchia* (*Platyhelminthes*) from sublittoral coastal areas near the Island of Sylt (North Sea). II. *Eukalyptorhynchia*. *Microfauna Marina* 5, 295-329.

Nematodes in Soils 18

DAVID J. HUNT AND PAUL DE LEY

International Institute of Parasitology, 395a Hatfield Road, St Albans, Herts AL4 0XU, UK.

Nematodes in Terrestrial Ecosystems

Nematodes exist as a diverse, highly speciated group in the soil environment and, although individually microscopic, often form the dominant biomass of the edaphic fauna. In addition to colonizing the soil they are also found in sediments and estuarine muds, often in colossal numbers. Many nematodes are microbivorous and feed on bacteria, yeasts and/or other fungal spores; others feed on the contents of fungal hyphae, root hairs or plant roots (by means of a hollow, protrusible stylet) or are predatory on other soil organisms or parasitic in arthropods, molluscs, annelids, etc. As a result they form an important component of the carbon and other nutrient cycles. In turn, they are preyed upon by fungi, mites, annelids, tardigrades and other nematodes. Although most species are vermiform and relatively mobile, many phytoparasitic species are endoparasitic or semi-endoparasitic in plant roots and may, in highly evolved forms, lose their motility in all stages except for the infective juvenile and the male, if present. The almost ubiquitous occurrence, diversity and abundance of the soil nematode fauna makes the group ideal as bioindicators of environmental stability and other factors such as pollution by heavy metal ions.

Sampling Strategy

The question of how much soil to sample is a difficult one to answer in terms of quantitative precision. Sampling strategy depends on the aim of the research and should be designed accordingly. This will profoundly influence the number and size of samples that are collected and whether the samples are taken at random or tied to a grid, for example. A further complication is the depth of sampling. Although most nematodes are found in the top 15-30 cm, some are found at much greater depths, particularly if the soil is dry. If a simple inventory of the nematode taxa is the main purpose, it is probably best to take a number of small (e.g. 200-500 ml volume) representative samples from the area to be investigated and then to bulk and thoroughly mix before subsampling. More complex ecological studies may demand individual extraction of each sample.

The majority of extraction techniques work best with soil volumes of 100-200 ml, although some can handle quantities in excess of 1 litre. If necessary, soils can be stored in polythene bags at about 5-10°C for several weeks without too many problems, although immediate extraction should always be the aim to avoid risk of subsequent alterations in the composition of the nematode fauna. Soils can often be extracted without further treatment, although if very stony they may be sieved through a suitably coarse mesh (e.g. 10 mm or more). Large lumps of soil may be gently crumbled before extraction or allowed to soak for a while to facilitate the process.

Extraction Techniques

The techniques for extracting nematodes from soil are many and varied. This is mainly a reflection of the unfortunate fact that no single technique is applicable to all groups of nematodes or all soil types. Extracting nematodes from soil is inevitably an operational compromise between the convenience or efficacy of the technique applied and the relative importance placed on the desired qualitative and quantitative parameters. Some techniques have the virtue of simplicity and may be performed away from the laboratory whereas others may require a combination of sophisticated equipment, a copious supply of clean, high-pressure water, a laboratory environment and skilled operators. This chapter aims to discuss the most common soil extraction techniques used in nematology and to indicate their advantages and disadvantages. It should be borne in mind, however, that, if a relatively complete inventory of the soil fauna is desired, then at least two techniques will need to be employed to approach such a requirement.

Nematode extraction techniques can be divided into two basic, although not necessarily exclusive, types, namely active and passive. Active techniques rely on the ability of the nematodes to migrate from the soil sample to the collecting vessel whereas passive techniques physically separate nematodes from the medium. It is therefore apparent that the former technique will discriminate against nematodes which are either quiescent, relatively immobile or present as non-motile stages such as eggs or cysts. Some techniques combine both strategies and initially employ a passive method to separate the nematodes from the bulk of the soil matrix followed by an active second stage to clean up the extract. This has the advantage that the movement of relatively immobile nematodes is facilitated by the removal of most of the soil particles and also by the consequent increased oxygen tension prevailing in the second stage. Many techniques require large volumes of clean water, perhaps at relatively high pressure, in order to operate as designed and this can be a severe constraint to an otherwise elegant and efficient methodology.

Active Methods

The main type of active technique involves separating the soil sample from the water in the collecting vessel by means of an extraction filter which allows passage of the nematodes and yet prevents the soil particles from contaminating and thus obscuring the extract. The most common choice of such a technique is one of the numerous variants of the 'tray method', itself an enhancement of the Baermann funnel technique (Fig. 18.1).

Baermann funnel

A typical Baermann funnel (Baermann, 1917) consists of a glass funnel with a piece of flexible tubing fitted with a clamp or screw clip attached to its outlet. This is mounted in a support on a stand and an extraction filter, composed of a porous gauze or paper filter supported by a nylon mesh, is placed inside. A thin layer of soil, cut-up leaves or other organic material is then spread out on the filter. Sufficient water should be added to the funnel so that the extraction filter plus material to be extracted is partly submerged. The set-up is then left to stand for one or more days. During this period, actively moving nematodes will crawl through the porous filter, sink down through the water in the funnel and collect in the flexible tubing above the clamp. They can then be removed by releasing a small amount of water through the clamp into a small beaker or dish.

The main advantages of the method are simplicity, minimal cost and the fact that the extracted nematodes end up in a minimal amount of water. Major disadvantages are: the relatively long running time; the danger of suffocating the animals; the inherent selection for more active species, stages and individuals; the inherent selection against species equipped with adhesive tail glands or hooks; and the unsuitability for quantitative analysis. The latter drawback is due to the many factors influencing efficiency and selectivity of each individual extraction, viz. sample size, shape and texture of each component of the set-up, temperature, running time, oxygenation, presence and interference of other soil organisms (especially predators). Several of these influences are difficult or impossible to control precisely, resulting in the lack of quantitative repeatability of the extraction.

Endless variations exist in the precise equipment and methodology used for 'Baermann funnel' extractions. Downwards movement of the nematodes can be stimulated by suspending a light bulb above the funnel, although this will select even more for very active animals and increase the risk of poor oxygenation. Oxygenation can be improved by adding a very small amount of H_2O_2 to the water, by using polyethylene tubing permeable to oxygen, or by modifying the set-up to a mist chamber extractor. In the latter case, the funnel is mounted under a sprinkler and above a washover beaker in a drained dish, the clamp being removed to allow a continuous, but gentle water flow.

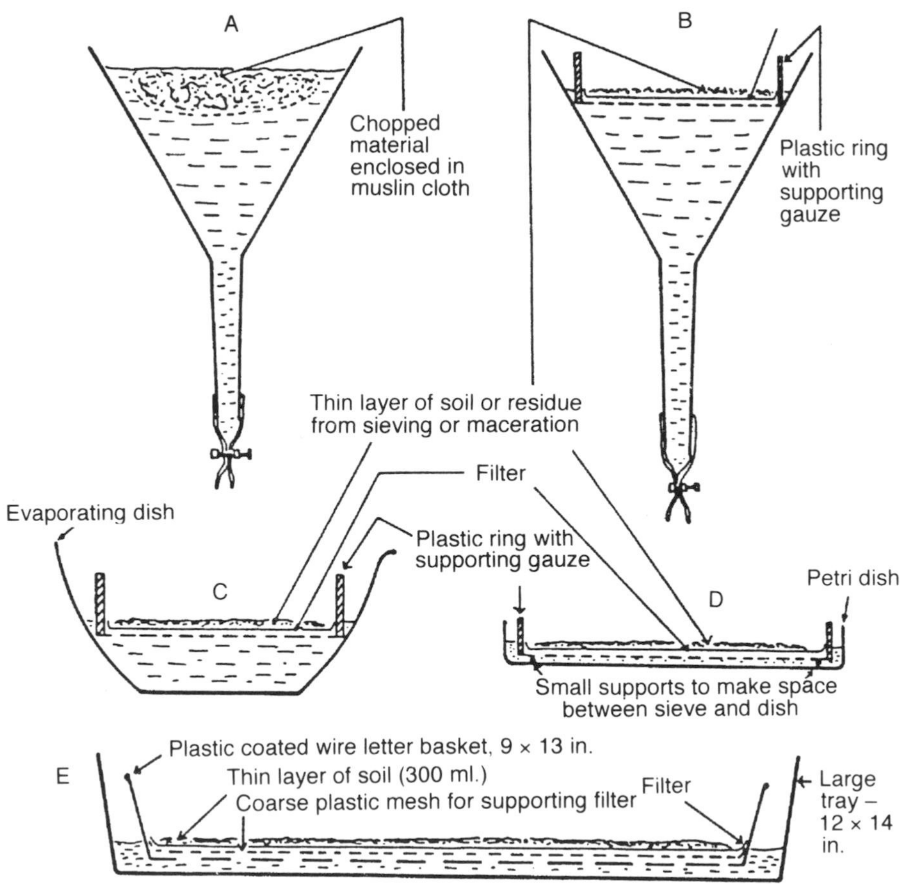

Fig. 18.1. Baermann funnel and modifications for extraction of active nematodes from chopped plant material, from thin layers of soil, or from residues obtained by sieving or maceration. Filter - cotton wool milk filter, 'wet-strength' facial tissue, coarse cotton cloth, or fine-woven nylon or 'Terylene' cloth. Plastic rings - cut from 'Perspex', polythene or vinyl tubes. Supporting gauze - muslin or nylon cloth held with an elastic band, coarse plastic mesh stuck or fused to edge of ring (after Hooper, 1990).

Some materials are best avoided if this technique is employed. The mesh supporting the filter should be made of nylon or stainless steel rather than brass, copper or other materials which may release sufficient toxic ions into the water to immobilize or kill certain groups of nematodes (e.g. *Dorylaimida*). A significant proportion of the nematodes will stick to the sides of the funnel and to minimize this effect the funnel should have steeply sloping sides and be made of glass or glazed ceramics rather than plastic.

Extraction tray

An extraction tray is an even simpler version of the Baermann funnel, utilizing a tray or dish to hold the porous filter and soil, instead of a funnel (Whitehead and Hemming, 1965). The tray, ideally made of plastic, can be adapted from a wide variety of cheap and locally available products. The filter, usually made from a single layer of facial tissue such as Kleenex®, must be separated from the bottom of the tray by means of a supportive mesh which prevents nematodes from moving back up through the filter once they have reached the bottom. Enough clean water should be carefully poured into the bottom of the tray such that the material to be extracted is moist, but not submerged. After one or two days, the mesh supporting filter and soil is carefully removed (avoid contaminating the extract with soil particles) and the water in the tray poured out. Care must be taken to agitate and/or rinse the tray sufficiently to detach as many nematodes from the bottom as possible.

This method has the advantages of flexibility and of avoiding oxygenation problems. The size of the tray governs the size of the sample which can be properly processed and, by using a large tray, samples of much greater volume (200 ml or more) can be treated than with the Baermann funnel technique (a large tray is also much cheaper and easier to find than a large funnel of appropriate quality). Also, most of the active nematodes will be extracted after two days. As a disadvantage, the nematodes end up in a larger volume of water and the extraction must be followed by one or more concentration steps in which excess water is removed.

Passive Methods

Passive methods are more diverse than active techniques and often exhibit considerable complexity, not to mention the higher cost and fragility of the apparatus. There are three major methods: *sieving*, *elutriation* and *centrifugation*. Sieving involves the use of a fine mesh to physically separate nematodes from the bulk of the soil matrix; elutriation involves separating nematodes from soil by utilizing the flow rate of an upcurrent of water, the strength of which is such that nematodes are held in suspension whilst the heavier soil particles sink, whereas centrifugation utilizes solutions of intermediate specific gravity to isolate nematodes from the soil particles. All methods involve fairly complicated apparatus and most depend upon a plentiful supply of clean water at the correct pressure. This can severely limit their usefulness, particularly under field conditions. Passive methods may make use of an active technique to finally clean up the extracted sample.

Sieving

Sieving (also known as decanting and sieving) involves pouring a suspension of soil in water through a bank of suitable sieves. Many modifications of Cobb's original technique (Cobb, 1918) have been made by subsequent workers and are described in more detailed publications such as that edited by Southey (1986). For most plant-parasitic nematodes sieve meshes of 75 µm and 53 µm are adequate and ideally there should, for maximum retention of nematodes, be two or three 53 µm sieves in a bank with one or two 75 µm sieves on top. A known volume of soil (200-400 ml) is placed in a large plastic bucket (about 2 litres capacity) and water added nearly to the top. Any large lumps of soil can be gently crumbled with the fingers and then the suspension is vigorously stirred for about 10 s and left for 30 s before being poured through the bank of previously moistened sieves. The material trapped on each sieve is then washed into a beaker with a gentle jet of water applied to the **back** of each sieve. The whole process can be repeated with the slurry left in the original bucket and the nematode suspension can either be examined directly or cleaned up by using one of the active methods outlined above.

This technique has the advantages of rapidly producing nematode extracts in a small volume of water and recovering immobile stages and/or large nematodes with reasonable efficiency. Disadvantages include the high initial cost of the precision sieves, the loss through the sieve mesh of a substantial proportion of the nematodes (this can be counteracted by having several consecutive sieves of the same mesh in the bank) and problems caused by a dirty extract when dealing with clay soils or those with a high organic matter content.

Immersion sieving

Immersion sieving is a variety of the standard sieving technique and is used for extracting long and delicate nematodes such as the longidorids (see Flegg, 1967). Immersion sieving should be regarded as supplementary to another extraction method. The set-up involves a 90 µm, or preferably 120 µm, sieve, because of the much larger nematodes to be extracted. The larger the sieve aperture, the cleaner is the extract, but this is achieved at the risk of losing most of the smaller nematodes through the mesh. A soil suspension is prepared as for the sieving technique and then poured through the 120 µm sieve, the mesh of which is just submerged by water in a plastic bowl modified by having overflow slots cut into the side at the appropriate level. After the suspension has been poured through, the sieve is gently washed with a jet of water to remove as much fine soil and debris as possible. The sieve is then removed from the bowl, and the residue washed off as described above.

This technique is efficient at recovering large (> 2 mm) nematodes such as longidorids and other dorylaims and reasonably effective at recovering small, less motile animals such as criconematids. A high proportion of the smaller soil nematodes will pass straight through the mesh, but this could be counteracted by using a finer-mesh sieve, something which is only practical in coarse, sandy soils which will not clog the sieve. Ideally, this method should be used as a supplement to a tray or funnel technique, the two combined giving a reasonably representative assessment of the nematode fauna.

Elutriation

Elutriation methods involve separating nematodes from the soil matrix by employing an upward flow of water, usually within a glass or Perspex® column. The separation relies upon the differing densities of soil and nematodes and the highly refined versions, by judicious selection of flow rate, can efficiently separate and hold in suspension nematodes of different sizes (Seinhorst, 1956, 1962). The apparatus designed by Seinhorst (1956, 1962) is too complex, fragile and expensive for general use and has fallen out of favour. It is not discussed further herein. The Oostenbrink elutriators (Oostenbrink, 1954, 1960) are more robust than the Seinhorst elutriator as they are made from metal rather than glass. They combine sieving with elutriation and can deal with relatively large samples of up to 1 litre. Banks of these elutriators can be assembled for efficient processing of large numbers of soil samples. Their main drawbacks are cost and the prodigious volume of clean water consumed during operation (see Hooper (1986) for an excellent account of their operation).

The Trudgill tower is a simple fluidizing column of robust construction and made from Perspex® (Trudgill *et al.*, 1973). Although it is often used solely for the extraction of cyst nematodes from soil, it can also be used for general soil nematodes, particularly in the extensively modified Wye Washer (Winfield *et al.*, 1987) version which is also capable of handling larger soil volumes (up to 1 kg) and extracting other soil invertebrates. Essentially, water of a controllable and variable flow rate is introduced to the base of a Perspex® cylinder which has a spout at the top end. The incoming water is dispersed evenly over the column diameter by means of a sintered glass or plastic plate. The substantial flow rate necessary should be controlled with a flow meter. The column is filled about one-third full with water and a soil sample of up to 200 ml added in the form of a slurry. The upcurrent can then be adjusted so that the nematodes and soil particles separate in the fluidized column for about 3 min. The flow rate is then increased to the appropriate level (i.e. about double) such that only the nematodes wash over the lip of the apparatus and are collected on a sieve or bank of sieves of appropriate mesh size. The flow rate varies depending on the size/type of nematodes to be extracted (see Hooper, 1990).

This apparatus is simple and robust, if expensive to make. It does require a supply of clean water under sufficient pressure to fluidize the column, but individual extractions can be completed in a matter of minutes. The Wye Washer modification is of considerably larger diameter and has a more efficient method for mixing the incoming water with the suspended soil in the column. It is more suitable for extracting vermiform soil nematodes and is less likely to clog.

Centrifugation

This approach relies on the use of differences in density to separate nematodes from denser soil particles and organic matter (see Caveness and Jenson, 1955; Jenkins, 1964; Hooper, 1986). The specific gravity of nematodes equals 1.15 g ml^{-1} on average, whereas soil particles usually exceed 1.20 g ml^{-1}. The advantages of centrifugation include the ability to extract not just the active nematodes, but also the inactive and dead ones; no requirement for pressurized water; its fast processing speed; the final concentration of the animals into a small volume of liquid; and the fairly good quantitative repeatability and taxa representation. Its disadvantages are the high cost of good equipment and separating solutions, failure to yield pure extracts from samples containing a large proportion of less dense particles (such as organic matter) and toxic and/or osmotic effects of many separating solutions, which reduce the viability and vitality of live nematodes.

A range of separating solutions can be used from simple chemicals, such as sucrose, $MgSO_4$ and $ZnSO_4$ solutions, to commercial products such as 'Ludox'® (colloidal silica), 'Ficoll'® (sucrose polymerized with epichlorohydrin) and 'Percoll'® (polyvinyl pyrolidone). Each of these has different effects on the viability, preservation quality and size classes of nematodes recovered, so that different solutions are optimal in different situations. $MgSO_4$ is usually preferred as giving the best compromise between cost and nematode survival, whereas the commercial products are usually quite expensive, but necessary if better survival is an important criterion. The speed and duration of centrifugation will also influence the composition and quality of the resulting extract.

Typical densities used for the separating solution range from 1.15 to 1.18 g ml^{-1}, and typical rotor speeds used are equivalent to 1800-2000 ***g***. Other densities and speeds can be used, but the practical range for density (1.15-1.23 g ml^{-1}) is far narrower than that for speed (700-2900 ***g***). In most centrifugation procedures, samples are manipulated prior and/or subsequent to the centrifugation itself, e.g. by sieving out large particles before spinning and washing away the separating solution and small low-density soil particles through a fine sieve after spinning.

Many modifications and refinements have been described in the literature and these are covered in greater detail by Hooper (1986). Bloemars and Hodda (1995) recently compared the efficiency of various combinations of different separation liquids, centrifugation speeds and centrifugation duration.

Specialized Extraction Procedures for Certain Taxa

Some specialized nematode groups live in the soil for only part of their life cycle. In many cases the soil phase of their life cycle is primarily a dispersal phase and the corresponding dispersal stage in their development cannot be adequately identified to species or even genus level. These nematodes require special approaches to either extract the identifiable adult stage, or to induce the dispersal stage to mature and reproduce in culture. An alternative is to apply molecular analysis to identify the soil-borne stages. The most important examples are:

Sedentary plant parasites

The adults of a number of groups are only found on or inside plant tissue, and the relevant tissues (roots, stems) must therefore be sampled to allow their identification. In the case of cyst-forming nematodes, special extraction devices are very useful to improve recovery of the mature female from soil, e.g. the Fenwick can (Fenwick, 1940) or Trudgill tower (Trudgill *et al.*, 1973). In the case of non-cyst-forming species, roots also need to be examined to obtain the mature females if specific identity is of importance. In all these nematodes, one (sometimes more) of the juvenile stages is to be found in the soil and can usually be identified to genus level with an acceptable degree of certainty.

Entomopathogenic nematodes

These are vectors of certain species of bacteria that are toxic to insects. The nematodes invade soil insects and release the symbiotic bacteria, which reproduce and cause a fatal septicaemia. The nematodes then feed on the resulting bacterial soup and reproduce prolifically before producing the infective stage, which migrates from the cadaver into the soil. The infective dispersal stages can be baited out of soil samples (Bedding and Akhurst, 1975; Fan and Hominick, 1991) and reared to the adult stage using wax-moth larvae (*Galleria*). A soil sample of 100-200 ml is sufficient for the baiting process, the live wax-moth caterpillars (or suitable alternative) being placed in the soil and left for several days before being recovered and observed for development of entomopathogenic nematodes. In tropical soils, ants and/or phorid flies can cause problems by consuming the wax-moth larvae and so care should be taken to exclude such unwelcome visitors.

Saprotrophic nematodes

A large number of soil nematodes feed on bacteria and a significant proportion of these are specialized for the invasion of microhabitats such as rotting wood, plant litter, dung, corpses, etc. Soil often only contains the dispersal stages of these species, which are usually impossible to identify precisely, even with molecular data (not due to methodology, but to the absence of a comprehensive database). They are often very ease to culture, however, and can be baited with rotting meat, a rotting potato or other decaying organic matter placed in a soil sample and left for a short period. The nematodes can then be extracted using one of the standard techniques.

Anhydrobiotic nematodes

Extraction of anhydrobiotic nematodes can cause problems if the active extraction methods are relied upon. Anhydrobiosis can occur in tropical soils during the dry season and in desert soils. Soaking the soil sample for a day or so may be sufficient to reactivate the nematodes sufficiently for tray or Baermann funnel methods to work, or they may be recovered with a centrifugation/ flotation technique (Freckman *et al.*, 1977).

Sample Concentration

The nematode suspension obtained from one of the tray or funnel techniques often needs to be concentrated before it can be processed further. The extract should be poured into a beaker or similar container and left undisturbed for several hours, in which time the nematodes should sink to the bottom. (N.B. There are some nematodes, such as *Aphelenchoides besseyi* or *Ditylenchus angustus*, which can actively swim and tend not to settle unless the suspension is cooled in a refrigerator at 4-10°C.) The surplus water can either be carefully decanted or siphoned off leaving the nematodes in a comparatively small volume of water. If a large initial volume of water is involved, the concentrating process may need to be repeated in several steps before a sufficiently small volume is achieved.

If the extract is reasonably clean, it can also be concentrated by pouring through a fine-aperture sieve (e.g. a 35 μm nylon or brass mesh), the nematodes then being carefully washed off into a small beaker or tube before being examined or killed and fixed. This method is rapid, but does tend to lose a significant proportion of the smaller nematodes and is thus more suitable if qualitative rather than quantitative assessments are the objective.

Killing and Fixing

Nematodes are best killed by gentle heat (i.e. 55-60°C) as this does not disrupt the body contents and specimens killed by this means assume a characteristic relaxed shape which can help in their identification. Nematodes can either be killed first and then fixed, or killed and fixed at the same time. The method chosen is usually down to personal preference. Soil samples can be fixed directly using one of the methods given below and the fixed nematodes subsequently recovered by a suitable passive extraction technique. **Correct fixation is essential for accurate identification of nematodes and care should be taken to see that the process is properly carried out. NEVER put live nematodes into cold fixative.**

Treating complete nematode suspensions is generally the most convenient method. After extraction, the nematode suspension is concentrated into a small volume of water (< 20 ml) by decanting, siphoning or centrifuging. This can be heated to a temperature of 55-60°C directly over a flame, on a hot-plate or in an oven; or by partially immersing the container with the nematode suspension in a large volume of water at approximately 80-90°C (off the boil) for a few minutes. Whichever technique is used the nematodes must **not** be boiled or overheated. After killing in this way the sample is left to cool and then the whole nematode suspension is fixed by adding an equal volume of 'double-strength' fixative. Alternatively, individual nematode specimens are picked out of the suspension and transferred into the cold fixatives at the normal strength shown below.

Fixing and killing of nematodes in suspension can be done using formaldehyde or FA 4:1. Double-strength fixative is heated to 80-90°C (just off the boil) and poured into an equal volume of nematode suspension, thus being diluted to the appropriate strength in the process.

Single or small numbers of nematodes can be killed and fixed in a drop of water on a glass slide in the same way as above, but overheating (and boiling!) is a real danger and should be avoided as it compromises identification.

The main fixatives used are TAF (Courtney *et al.*, 1955), FA 4:1 (Seinhorst, 1966) and dilute formalin. The formulae for FA 4:1 and dilute formalin may be modified by the addition of 1% glycerol to avoid danger of the extract drying out.

Fixative Formulations

	Normal strength	Double strength
TAF		
40% Formaldehyde (formalin)	7 ml	7 ml
Triethanolamine	2 ml	2 ml
Distilled water	91 ml	45 ml
FA 4:1		
40% Formaldehyde (formalin)	10 ml	10 ml
Glacial acetic acid	1 ml	1 ml
Distilled water	89 ml	45 ml
Formaldehyde (formalin)	2% (5%)	4% (10%)

Nematodes in fixative should be left for 12 h or overnight before processing further.

CAUTION! These solutions contain hazardous compounds and appropriate safety precautions must be followed.

Limitations of Existing Techniques

The main problem with extracting nematodes from soil is that nematodes are so diverse in their biology and morphology that no single technique is capable of providing a representative qualitative and quantitative sample of the entire fauna. This problem is exacerbated in that soils vary tremendously in their composition and such factors as clay content or presence of high organic matter levels, for example, seriously affect extraction efficiency. These limitations can be partially overcome by employing a minimum of two different extraction methods and combining the results, although certain aspects of the nematode fauna will still be discriminated against and be under-represented. A further problem is the requirement of many techniques for a copious supply of clean, high-pressure water to facilitate the separation of the nematodes from the soil matrix. Although this may not be a problem in a well-serviced laboratory situation, it does pose considerable difficulties in more remote locations.

Nematode Assessment as Part of a General Protocol

Nematode assessment is best carried out by dedicating at least one small subsample (e.g. 100-200 ml) from each sample or bulk sample collected. For a reasonably comprehensive inventory, two such samples should be extracted using complementary techniques and at least 200 specimens should be randomly picked from each and mounted on slides for identification. Identification to genus level may be sufficient for ecological studies and usually takes little time as it can be done 'on sight' under a high-power microscope. Identification to species level is usually more time-consuming as it often involves taking measurements from each specimen.

Certain other taxa can be recovered simultaneously with nematodes from soil subsamples. Although active extraction techniques often yield live ciliates, enchytraeid worms, turbellaria, rotifers and tardigrades, the relative recovery efficiency for such groups is unknown, but likely to be low in the case of tardigrades, for example. In addition, soil or sediment used for extraction of nematodes is usually recoverable from the extraction filter and, if necessary, could subsequently be utilized for isolating other groups of soil organism which would not have been adversely affected by the extraction process. For 'All Taxa Inventory' purposes, however, it will usually be more efficient to simply use separate subsamples for recovering taxa that require significantly different extraction methodologies from *Nematoda*.

Bibliography

General Texts

Hooper, D.J. (1990) Extraction and processing of plant and soil nematodes. In: Luc, M., Sikora, R.A. and Bridge, J. (eds) *Plant Parasitic Nematodes in Subtropical and Tropical Agriculture.* CAB INTERNATIONAL, Wallingford, pp. 45-68. [General overview of extraction and processing techniques.]

Hooper, D.J. and Evans, K. (1993) Extraction, identification and control of plant parasitic nematodes. In: Evans, K., Trudgill, D.L. and Webster, J.M. (eds) *Plant Parasitic Nematodes in Temperate Agriculture.* CAB INTERNATIONAL, Wallingford, pp. 1-59. [General overview of extraction techniques.]

Oostenbrink, M. (1960) Estimating nematode populations by some selected methods. In: Sasser, J.N. and Jenkins, W.R. (eds) *Nematology.* University of North Carolina Press, Chapel Hill, pp. 85-102. [Comparison of extraction techniques.]

Southey, J.F. (ed.) (1986) *Laboratory Methods for Work with Plant and Soil Nematodes. Reference Book 402.* HMSO, London. [Excellent manual for nematology techniques.]

Specific Texts

Baermann, G. (1917) Eine einfach Methode zur Auffindung von *Ankylostomum* (Nematoden) Larven in Erdproben. *Geneesk.undische Tijdschriften Nederlandische-Indië* 57, 131-137.

Bedding, R.A. and Akhurst, R.J. (1975) A simple technique for the detection of insect parasitic rhabditid nematodes in soil. *Nematologica* 21, 109-110.

Bloemars, G.F. and Hodda, M. (1995) A method for extracting nematodes from tropical forest soil. *Pedobiologia* 39, 331-343.

Caveness, F.E. and Jenson, H.J. (1955) Modification of the centrifugal-flotation technique for the isolation and concentration of nematodes and their eggs from soil and plant tissue. *Proceedings of the Helminthological Society of Washington* 22, 87-89.

Cobb, N.A. (1918) *Estimating the Nema Population of the Soil.* Agricultural Technology Circular No. 1, Bureau of Plant Industry, US Department of Agriculture.

Courtney, W.D., Polley, D. and Miller, V.L. (1955) TAF, an improved fixative in nematode technique. *Plant Disease Reporter* 39, 570-571.

Fan, X. and Hominick, W.M. (1991) Efficiency of the *Galleria* (wax moth) baiting technique for recovering infective stages of entomopathogenic rhabditids (*Steinernematidae* and *Heterorhabditidae*) from sand and soil. *Revue de Nématologie* 14, 381-387.

Fenwick, D.W. (1940) Methods for the recovery and counting of cysts of *Heterodera schachtii* from soil. *Journal of Helminthology* 18, 155-172.

Flegg, J.J.M. (1967) Extraction of *Xiphinema* and *Longidorus* species from soil by a modification of Cobb's decanting and sieving technique. *Annals of Applied Biology* 60, 429-437.

Freckman, D.W., Kaplan, D.T. and Van Gundy, S.D. (1977) A comparison of techniques for extraction and study of anhydrobiotic nematodes from dry soils. *Journal of Nematology* 9, 176-181.

Hooper, D.J. (1986) Extraction of free-living stages from soil. In: Southey, J.F. (ed.) *Laboratory Methods for Work with Plant and Soil Nematodes.* Reference Book 402, HMSO, London, pp. 5-30.

Jenkins, W.R. (1964) A rapid centrifugal-flotation technique for separating nematodes from soil. *Plant Disease Reporter* 48, 692.

Oostenbrink, M. (1970) Comparison of techniques for population estimation of soil and plant nematodes. In: Phillipson, J. (ed.) *Methods of Study in Soil Ecology.* UNESCO, Paris, pp. 249-255.

Rodriguez-Kabana, R. and King, P.S. (1975) Efficiency of extraction of nematodes by flotation-sieving using molasses and sugar and by elutriation. *Journal of Nematology* 7, 54-59.

Seinhorst, J.W. (1956) The quantitative extraction of nematodes in soil. *Nematologica* 1, 249-267.

Seinhorst, J.W. (1962) Modifications of the elutriation method for extracting nematodes from soil. *Nematologica* 8, 117-128.

Seinhorst, J.W. (1966) Killing nematodes for taxonomic study with hot f.a. 4:1. *Nematologica* 12, 178.

Trudgill, D.L., Evans, K. and Faulkner, G. (1973) A fluidising column for extracting nematodes from soil. *Nematologica* 18, (1972), 469-475.

Whitehead, A.G. and Hemming, J.R. (1965) A comparison of some quantitative methods of extracting small vermiform nematodes from soil. *Annals of Applied Biology* 55, 25-38.

Winfield, A.L., Enfield, M.A. and Foreman, J.H. (1987) A column elutriator for extracting cyst nematodes and other small soil invertebrates from soil samples. *Annals of Applied Biology* 111, 223-231.

Land and Freshwater Molluscs and Crustaceans 19

Mary B. Seddon and P. Graham Oliver

Department of Zoology, National Museum of Wales, Cathays Park, Cardiff CF1 3NP, UK.

Molluscs in Terrestrial Ecosystems

Land molluscs are estimated to comprise some 35 000 species. Most molluscs lay their eggs in the soil and so they are commonly found in the upper parts of the soil and in the litter layer. They may also have a role in incorporating organic material into the mineral layers of the soil. Deep soil-dwelling species are relatively few. Molluscs are primarily detritivores or microherbivores with only a few carnivorous species. They are important in the decomposer cycle and are, in turn, preyed upon by a variety of insects and all vertebrate groups.

Endemism of non-marine molluscs is high; the median range has been estimated to be less than 50 km^{-2} (Solem, 1984). The highest levels of endemism are found on islands and in isolated mountain ranges. Comparisons between molluscan diversity in different continents and different climatic zones are still some way off, as many tropical areas are inadequately sampled, especially for the litter element. However, recent work on molluscan biodiversity using a quantitative approach has revealed interesting patterns in regional diversity (Tattersfield, 1989; Cameron, 1995; Cowie, 1995). This work also showed that it is possible to identify differences in natural and disturbed habitats; hence molluscan biodiversity may be a useful indicator of the degree of stress the ecosystem is suffering.

The type of geology and landscape influence both the species diversity and population densities, the highest numbers being found where the ground is calcareous (e.g. limestone, calcareous sandstone). Precipitation rates also influence the species diversity; arid and semi-arid areas usually having a lower diversity than temperate or tropical regions. Soil structure and chemical structure of the litter and soil also appears to play a role in the number of species present in the soil (Outeiro *et al.*, 1993).

Site Selection

The sampling interval will vary depending on accessibility of terrain, size of survey area and the nature of the survey (whether qualitative or quantative).

Regional Level Surveys

Two types of survey are commonly undertaken:

Grid mapping. Recent country surveys used a grid mapping approach with squares of 50, 10 and 2 km^2. In each square, a variety of microhabitats are sampled, and the data held at grid square level.

Point mapping. Older studies used point sampling rather than grid mapping. These techniques tend to be more applicable in less accessible terrain and those with large altitudinal ranges.

- Recommended intervals for site separation when surveying molluscs is less than 30 km.
- Select sites which represent the diversity of microhabitats, e.g. rocky slopes with low crags, springs and streams, native woodland, grassland, marshes, etc.
- Care has to be taken to find sites of primary habitats rather than disturbed habitats, which may contain introduced species, e.g. on logging tracks in native forest, secondary rather than primary forest habitats may be accessed.

Area Surveys

Surveys of smaller regions or specific habitats allow a different strategy to be taken.

- Timed searches at specific points (radius of search = 10 m) along transects crossing the area (Cowie, 1995).
- Timed searches in areal plots (e.g. 40 m^2 in Tattersfield, 1989) selected according to vegetation type.
- Quadrat surveys of specific areas (25 cm^2 in Young and Evans, 1991; 50 cm^2 in Outeiro *et al.*, 1993).

Sampling Methods

Sampling for terrestrial molluscs requires a range of techniques to obtain the entire fauna, due to the lifestyles and sizes of different taxa. The slow-moving nature of molluscs renders them unsuitable to the behavioural extraction methods suitable for arthropods. Pitfall traps in particular do not give a representative sample of the molluscs present in an area. Many molluscs are nocturnal in behaviour, and so the sampling techniques are directed to their resting sites. If, however, it has recently rained, land snails are far more active, and start to move onto surfaces such as tree trunks, plant stems or rock-faces to feed. Many of the detritus-feeding litter snails are small (under 4 mm in diameter) and, as a consequence, tend to be missed unless systematic sieving is used for sampling. These factors must be taken into account when comparative work on diversity at sites is considered.

In some environments, the seasonality of the climate means that the mollusc aestivates through the dry season, and, in these cases, the land snail may move to depths of several metres into the soil through vertical cracks. Other aestivation sites include deep rock fissures, the undersides of large boulders and trees or posts. Slugs tend to creep into these fissures more rapidly than other land snails, and so they should be searched for in moister microhabitats.

The shell of molluscs may remain for long periods after the animal has died, the length of time it persists being related to the acidity of the leaf litter and precipitation rates. The most rapid rates of shell decay are apparent in tropical forest environments (< 10 days), whereas, in arid environments, breakdown may take centuries. This may mean that the shells need to be classified when analysing the data (e.g. live collected, dead shells with periostracum, worn dead shells, subfossil shells).

Sampling for Molluscan Biodiversity

At each site, the same techniques should be utilized in order to obtain comparable data for qualitative or quantitative work. This should combine direct search methods for the larger species and collection of litter and soil samples for processing for the 'micro-species'. Be careful to minimize the disturbance of the habitat, as sampling may be more destructive when collecting a rare species.

Care should be taken to select a suitable range of microhabitats. Fringes of habitats often have a more diverse molluscan fauna than the central part of the habitat. The reasons may be related to introduced species, but more often reflect differences in the amount of canopy shade/proportion of light reaching the forest floor and the diversity of its flora.

Direct search

Larger species (including slugs) tend to be found easily by direct searching of the ground surface and leaf litter. Areas where concentrations of shells are more easily found are litter/soil at the base of trees, base of rock crags, and in damp areas around springs. Stones should be rolled over and fallen wood should be examined where lying on the leaf litter. If undertaking quantitative work, use a fixed time and area.

Smaller species can be found by shaking litter from the base of crags, trees and grasses over a white tray. Otherwise use the sieving techniques which follow.

Samples collected should be placed into appropriate containers, labelled adequately (outside and inside) and, if time is available, fixation of the material for future work should be started.

Leaf and soil litter

Sieving of leaf litter and soil is generally used both for qualitative and quantitative analysis of land-snail diversity (e.g. Kerney and Cameron 1979; Cowie, 1995). There is variability in the impact of sieving on the number of species recorded at any one site. This tends to relate to the percentage of small species in the fauna, as well as the experience of the collectors and the number of collectors.

Typically litter/soil samples are taken from an area 0.25-0.50 m^2. The sample can be subdivided if the volume of molluscs is too high. The amount of leaf litter to be processed decreases in regions with low litter levels, but 0.5 m^2 in tropical forest can provide up to 4 litres.

In the field, samples should be sieved coarsely (to about 4 mm) and the large shells picked from these samples by hand. The residue should then be removed for processing. The most useful sample bags are thin cotton bags or fine-mesh bags, as these ensure that damp litter can dry rather than starting to decompose, so that live animals survive. The sample is easier to sort when dry (preferably air-dried, so that the species remain alive). It should be sieved again (2 mm and 0.5 mm sieves) to remove the fine dust prior to picking. Wet sieving and flotation can be used (with caution), but may lead to difficulty picking samples, and the small mesh size required to get the small molluscs often results in clogging of the sieve (Wallwork, 1970).

Each residue is then sorted on trays (under low magnification). Shells should be picked up with either a small paintbrush or **soft** forceps. Live material should be separated from dead shells, and killed for identification.

Preservation of Material

Samples should be processed within 5 to 6 hours, or less in hot climates. This will ensure that the live material can be fixed for taxonomic work. Some species cannot be identified on the basis of the shell only and hence the bodies should be preserved to allow examination of the anatomy. There is (as yet) no reliable method for killing and fixing pulmonate land snails in an extended state. Different fixing techniques need to be employed for molecular systematics.

Preservation of terrestrial molluscs

Formalin should **not** be used for any long-term storage of material as it leads to deterioration and cracking of the shell, hardens the muscular body wall and makes the soft parts brittle. The body is thus difficult to use for anatomical work. Ideally, material should be drowned (in a relaxed state) and then properly fixed. Most collections are stored in 70% (v/v) alcohol (sometimes with a small % of glycerol).

Large species of land snail

The success of the technique will vary with the taxonomic group and the field conditions. The best method for killing the animal is to drown it by leaving it in a full, sealed tube of deoxygenated water (freshly boiled and cooled water) for 8 to 12 hours. It is also possibly to drop the specimen in boiling water, but this requires skill not to overcook the species. It is also possible to use a weak solution of chloral hydrate or nicotine.

Once the species has drowned, the killing medium should be replaced with 40% (v/v) alcohol. After a few hours, the alcohol should be changed, increasing its concentration to 60% (v/v), with a subsequent change to 70% (v/v). On return to long-term storage, change the alcohol to ensure that the specimen is completely fixed.

Shells which are mainly periostracum should always be stored wet, even if the bodies are not preserved.

Slugs

The best method for killing the animal is to drown it by leaving it in a full sealed tube of deoxygenated water (freshly boiled and cooled water) with a few drops of alcohol for 6 to 12 hours (less in hot climates). Once the species has drowned the water should be replaced with 30% (v/v) alcohol. After a few hours, change the alcohol, increasing to 40% (v/v), with subsequent changes to 60% (v/v) and 70% (v/v). On return to long-term storage, the alcohol should be changed to ensure that the specimen is completely fixed.

Small species of land snail (under 3 mm)

It is more difficult to get relaxed specimens of small species. One method is to drown the snails as outlined above. Decomposition occurs more rapidly so fixing of the material has to be done quickly. An alternative way to ensure that the animal is relaxed is to place the animal in a small glass tube. Allow the animal to start crawling around the tube, then dip the tube quickly into near-boiling water. This kills the animal and then the fixing process should be followed, with alcohol added at 40% (v/v), then 70% (v/v), with a time interval dependent on the rate of decomposition and fixing of alcohol in the body.

Equipment needed

Field equipment

Set of sediment sieves (e.g. Endicott sieves: 4-5 mm, 2-3 mm, 0.5 mm)
White plastic trays (*ca.* 40 cm × 25 cm)
Mesh (0.5 mm) bags for litter samples
Storkbill forceps
Paintbrushes
Hand-lens (× 10)

The Importance of Molluscs in Freshwater Ecosystems

In faunal freshwater molluscs are limited mostly to the bivalve superfamilies *Unionoidea* and *Corbiculoidea*, with most freshwater gastropods living with macrophytic plants. Freshwater bivalves are primarily suspension feeders and play a key role in maintaining water purity. Smaller species are important sources of food for fishes. Unionid clams have undergone regional diversification and there is a high degree of endemism among them.

Freshwater sediments can be sampled, fixed and preserved in a similar way to that described for marine benthic macrofauna and the same considerations described there apply (Chapter 21). Generally, the scale of the operation is smaller because many freshwater bodies are small and unsuited to working from seagoing vessels. The initial sampling can be done using small versions of the Petersen and Ekman type of grab, or by coring. A large variety of cores have been developed, each specific to the substrate, size and depth of the water body involved. These are reviewed by Hynes (1971) and Southwood (1976). Diver-operated cores are probably the most consistent, but may not always be practical due to safety considerations. In streams, especially if the sediments are coarse, the 'kick sampling' method is still widely used. Quite simply, this involves disturbing a known area of stream bed and allowing the mobile fauna to drift downstream into a net.

Sampling of the large freshwater mussels in any destructive way is now to be discouraged because of their sensitivity to interference. Dredges and grappling hooks were used commercially in the large North American river systems and could be used in areas where stress on the populations would not be incurred. Direct observation or use of SCUBA in deeper waters is probably to be recommended, depending on safety factors.

The sediment samples should be wet-sieved through a 0.5 mm or 1 mm mesh and the fauna removed using a low-power dissecting microscope. There are a number of methods for relaxing freshwater molluscs (Araujo *et al.*, 1995) and the fixative of choice is 80% (v/v) ethanol.

Terrestrial and Freshwater Crustaceans

Terrestrial crustacea can be sampled in identical ways to soil- and ground-dwelling insects and arthropods (see this volume, Chapter 13) and include the following methods: direct searching of the habitat or resting places by searching under stones, logs or in litter; pitfall trapping for mobile surface or litter species; behavioural techniques based on the Tullgren funnel, wet sieving or flotation for soil and litter species.

Benthic freshwater crustacea can be sampled following a similar protocol to their marine counterparts (see this volume, Chapter 15) but with modifications controlled by the size, depth and flow rate of the water mass being surveyed. The methods outlined for freshwater mollusca can also be applied.

Bibliography

Araujo, R., Remon, J.M., Moreno, D. and Ramos, M.A. (1995) Relaxing techniques for freshwater molluscs: trials of different techniques. *Malacologia* 36, 29-42.

Cameron, R.A.D. (1995) Patterns of diversity in land-snails: the effects of environmental history. In: van Bruggen, A.C. and Wells, S. (eds) *Diversity and Conservation of Molluscs*. E.J. Brill/W. Backhuys, Leiden (in press).

Cowie, R. (1995) Patterns of land-snail diversity in a montane habitat on the island of Hawaii. *Malacologia* 36, 155-169.

Groombridge, B. (ed.) (1992) *Global Biodiversity. Status of the Earth's Living Resources.* Chapman and Hall, London.

Hynes, H.B.N. (1971) *The Ecology of Running Waters*. Liverpool University Press, Liverpool.

Kerney, M.P. and Cameron, R.A.D. (1979) *A Field Guide to the Land Snails of Britain and North-West Europe.* Collins, London.

Outeiro, A., Aguera D. and Parejo, C. (1993) Use of ecological profiles and canonical corrspondence analysis in a study of the relationship of terrestrial gastropods and environmental factors. *Journal of Conchology* 34, 365-375.

Solem, A. (1984) A world model of land-snail diversity In: van Bruggen, A.C. and Wells, S. (eds) *World-wide Snails. Biogeographical Studies on Non-Marine Mollusca.* E.J. Brill/W. Backhuys, Leiden, pp. 6-22.

Southwood, T.R.E. (1976) *Ecological Methods*, 2nd edn. Chapman and Hall, London.

Tattersfield, P. (1989) Terrestrial mollusc faunas from some South Pennine woodlands. *Journal of Conchology* 33, 355-374.

Wallwork, J.A. (1970) *Ecology of Soil Animals.* McGraw-Hill, London.

Williamson, M.H. (1959) The separation of molluscs from woodland leaf-litter, *Journal of Animal Ecology* 28, 153-155.

Young, M.S. and Evans, J. (1991) Modern land mollusc communities from Flatholm, south Glamorgan. *Journal of Conchology* 34, 63-70.

Earthworms 20

Sam W. James

Maharishi International University, Fairfield, Iowa 52557-1056, USA.

The Importance of Earthworms in Ecosystem Processes

Earthworms (*Annelida: Oligochaeta*) are found in most soils of temperate and tropical ecosystems, except in arid to semi-arid regions. They occur in boreal forest to a limited extent, but mostly as introduced species in anthropogenic habitats. In general, wherever there is sufficient rainfall to keep the soil moist for three to four months of the year, earthworms can be found. However, given the extensive modifications of soils and the tremendous alterations of vegetation caused by human activity, earthworms may be absent from areas where one would expect to find them, or present where one would not, such as in irrigated soils of desert regions. Furthermore, numerous earthworms have achieved cosmopolitan distributions through human transport of soils and plants. These peregrine earthworm species may replace an indigenous earthworm fauna, usually with human-caused habitat destruction as an aid, or they may co-occur with indigenous earthworm species. On some oceanic islands the presence of earthworms may be due entirely to importation of peregrine species (Fragoso *et al.*, 1995).

Earthworms are generally among the largest contributors to soil fauna biomass and are important for their contributions to organic matter decomposition and comminution, nutrient mineralization, soil profile development and soil macropore development (Lee, 1985; Swift and Anderson, 1989; James, 1991; Blair *et al.*, 1995). Earthworm biomass routinely exceeds the total above-ground vertebrate biomass supported on the same site. In Puerto Rican humid forest, earthworms comprised 30% of the total faunal biomass of the forest (Moore and Burns, 1970). There is a considerable range of abundance in natural systems, and contribution to functions of the ecosystems will vary accordingly. Even though earthworms are known as decomposers, earthworms can protect organic matter from further breakdown in humid tropical systems (Lavelle and Martin, 1992). Where this is true, earthworms are helping to create soil carbon sinks, an important ecosystem service with implications for global atmospheric composition and for soil quality. An inventory of earthworms will provide valuable preliminary information that can be used by anyone interested in pursuing these questions further.

In addition to their impact on soils and on biogeochemistry, earthworms are an important source of food for other organisms. Numerous small mammals of South-East Asian and Madagascarian forests appear specialized to feed on earthworms and other soil invertebrates. More generally, many insectivore mammals and some birds depend in part on earthworms as food. Human use of earthworms as food has been documented from indigenous cultures of at least three continents.

Sampling Techniques

Seasonality of Collecting

In the tropics, collecting in the wet season is recommended. Earthworms will be in the adult stage (this is critical to identification) at some point during the wet season, and will of course be most active and accessible at this time of year. At present there is not enough known of the phenology of earthworm reproduction and life cycles in most tropical regions to allow prediction of the best part of the rainy season for any particular species. For a first approximation, a good time might be one month after the beginning of the rainy period, when other more visible organisms have clearly decided to respond to the rain. It may be that different species come into maturity (and hence can be identified or described) at different times. The second approximation would be to collect the same sites in the late rainy season, or after the short dry period (if there is one) near the solstice.

Mediterranean climate regions should be collected during the winter rainy period at a time when the temperatures are warm enough for earthworm activity (above 10°C). In cooler temperate regions sampling should be primarily in the spring and into early summer, provided soil temperatures are not extremely high (28-30°C) and soil moisture is not depleted. Autumn may provide additional sampling time.

Extraction Techniques

Earthworm collecting is a very simple but highly seasonal enterprise. Given that the likely places are known to the collector, more depends on persistence and timing than on finesse or technological sophistication. Methods for obtaining earthworms from soil samples rely on physical means of extraction, chemical expellants, vibrational expulsion and electrical currents. Physical or mechanical techniques are the most effective on the vast majority of earthworm species. Other methods are resorted to when these fail, or when, in the case of commercial collections, large numbers are desired in a short time. All of the non-physical methods rely on the behaviour of the earthworms in response to some stimulus. It should be kept in mind that these responses vary among taxonomic groups and among ecological niche categories of earthworms. Furthermore, the responses are not known for the vast majority of species.

Physical extraction

The simplest mechanical method is digging and hand-sorting soil samples. This is, indeed, the method most persons would spontaneously choose if told to go and find earthworms. It is applicable in any soil type, though it can be difficult in rocky or densely rooted soils. It has the advantages of relying on the senses of the collector, and not on the behaviour of the animals in response to some stimulus. Results are apparent very quickly, the equipment required is universally available, and the whole procedure is portable to a degree not always true of the other methods.

More efficient recovery of earthworms from soils can be achieved by wet-sieving/flotation techniques (Andrews, 1972), which rely on earthworms floating on a solution of magnesium sulphate. Any wet-sieving technique will require more hardware than digging, and a significant amount of water. Wet-sieving and flotation are best used for very accurate quantitative studies of earthworm population dynamics, and may not be appropriate for a simple inventory and first approximation of populations.

Limitations of digging and hand-sorting soil include the labour-intensive nature of the work, particularly when care is taken to remove every earthworm from the soil sample. Also, it is not very effective at recovering deep-burrowing species, since the amount of physical effort seems to grow exponentially with depth of hole to be excavated. Wet-sieving and flotation techniques are more capital-intensive, and render the soil sample unusable for the extraction of other biota. Sieving is difficult with well-aggregated or otherwise cohesive fine-textured soils and with soils high in organic matter. In the case of the latter difficulty, worms can be hidden by the organic material with which they float out. In such cases, sieving becomes very labour-intensive.

Behavioural techniques

Earthworms have been recovered from soil and litter samples taken for other purposes, and placed in Berlese funnel apparatus (Kempson *et al.*, 1963). Like other elements of the soil fauna, the earthworms avoid heat, dryness and light. This method may be good for small species dwelling in litter where they are hard to find. For routine collecting, the volume of soil that would have to be placed in funnels might be prohibitive.

Chemical expellants

These have been used routinely on the *Lumbricidae*, a Holarctic (but predominantly European) family with many peregrine species now domiciled in other temperate regions and in high-elevation tropical systems. The chemical most often used is formaldehyde, at 0.4% (v/v) concentration in water (Raw, 1959), but chloroacetophenone is also effective (Daniel, 1992). Typically an enclosure, such as a sheet-metal ring, is inserted a short distance into the soil to serve as a dam, and the formaldehyde solution is poured into the enclosure. After some 5-15 min, the solution penetrates to the earthworm burrows and they come up to the surface of the soil. The advantages are that no digging is necessary and that deep-dwelling species can be easily obtained from their burrows. If the animals are quickly collected and rinsed in fresh water, they may survive; otherwise they will soon be narcotized and will die. This method is astonishingly effective when it works.

There are other chemicals usable as expellants, though none is so well documented as formalin, and the taxonomic breadth of application is not known for any. Most of these chemicals have emerged from a sort of folklore of earthworm collecting that has arisen from the observations of fishermen and landscape gardeners. Describing his use of iron sulphate, Eisen (1900) states ‘A teaspoonful of the sulphate dissolved in a bucket of water and poured over the ground will soon cause the worms to make their appearance.’ Dry mustard powder, the hotter the better, can be mixed with water and applied to the soil. Aqueous extracts of the seed of the neem tree (*Azadirachta indica*), native to the Indian subcontinent, and now used in several commercially available insecticide formulations, have been reported to expel worms. Calcium polysulphide, also known as lime sulphur, is used by bait collectors on the megascolecid North American genus *Diplocardia*. Two insecticides, 'Mocap' (Rhône-Poulenc Co.: *o*-ethyl *s,s*-dipropylphosphorodithioate) and 'Dursban' (Dow-Elanco: *o,o*-diethyl *o*-(3,5,6-trichloro-2-pyridinyl) are also known to expel *Diplocardia* species from the soil. All of these alternatives to formaldehyde will need testing of the concentrations of the chemical as well as determination of the taxonomic range of effectiveness, since no data are available. The folklore is not finely tuned.

Vibrational techniques

Another method of extracting earthworms from soils in the field relies on earthworm response to vibrations. It is often observed that when digging in one spot, earthworms are found fleeing onto the soil surface some small distance away. Again, the species which respond in this manner are likely not to be the entire set inhabiting the site, and there may be taxonomic differences here as well. Commercial bait collectors in the southern and central USA use two methods. The first (used primarily in Florida and Georgia) is to drive a slat of wood (approximately 1.5 cm × 4 cm × 1 m) into the ground until it is stiffly anchored. A piece of rough-sawn lumber, an old car leaf spring or other suitably coarse-textured object is then rubbed across the top of the slat to make a low vibration. The method is known as 'grunting', to give an idea of the sound to be produced. This is kept up for several minutes and worms emerge within a few metres of the stake. A second method more common in the central states of Missouri and Kansas, and perhaps elsewhere, is to use a gasoline-powered soil tamper, a machine resembling a lawnmower but which delivers soil-tamping impacts to the ground. One person slowly moves the tamper while others collect earthworms off the surface.

Vibrations also work on *Megascolecidae* of California, where Eisen (1900) used a 2.5 cm thick metal rod driven into the ground and worked sideways for a few minutes.

Vibrational techniques are used to collect large-bodied earthworms that feed at or near the soil surface. Other species may respond to the vibrations but, since they are not commercially valued as bait, it is not known if they also emerge. In the author's experience, vibration is effective on many small and large *Diplocardia* species of the southern USA, but it is not known whether this method is useful on earthworms of other regions. Finally, this approach to earthworm collecting is not quantitative, and like chemical extraction depends on the activity status of the earthworms and the soil moisture conditions.

Electrical shocking

Electrical shocks are a relatively non-invasive method of removing worms from soil in the field (Bohlen *et al.*, 1995). A simple and 'traditional' method is to connect the terminals of an automobile battery to metal rods driven into the ground. Like the chemical and vibrational methods, there is some taxonomic variability in effectiveness, and it depends on soil conditions and the activity status of the worms. It is not known if earthworms flee the electrical field in all directions or only by going up.

Rainfall

Nature provides a method to collect worms without digging, chemicals or cumbersome equipment, and it seems to be very broad, perhaps even universal, in its taxonomic range of application. Even in tropical rain forests, soil-dwelling earthworms will emerge during heavy rains. Thus, if one does not mind going out into the field during and immediately after heavy, prolonged rainfall, earthworms can be collected easily. They may be hiding under leaf litter or obscured by low herbaceous vegetation, but it does not take long to search a considerable area. This may be the best way to obtain complete specimens of large earthworms normally inaccessible to digging and unresponsive to chemicals and vibrations. The only disadvantage is the unpredictability of the required rainfall. This method may be best left to those who inhabit the area from which earthworms are desired, and can rise to the occasion.

Earthworms from Habitats Other Than Ordinary Soil

Not all earthworm species inhabit ordinary soils, by which is meant soils not periodically submerged, chronically saturated with water or extremely high in organic matter (histisols). Some species do not live in soils at all. The purpose of this section is to alert the reader to these other possibilities and to emphasize what appear to be the best ways to obtain specimens of earthworms from unconventional habitats.

There are many mud-dwelling earthworms, including a few families that are almost entirely confined to chronically wet, low-lying soils or sediments, such as those of riparian areas and near other bodies of water or in wetlands. Penetration of expellant chemicals will be very slow, relying as it must on diffusion rather than on percolation of the solutions through soil pore spaces. Spontaneous emergence during heavy rainfall is unlikely, and the efficacy of vibration and electrical shocking are unknown. Digging is probably the most effective method, though it is certain to be laborious. *Sparganophilus eiseni* (*Sparganophilidae*) has been found in mud at the water's edge and in lake bottom sediments (Smith and Green, 1916). *Eisenoides lonnbergi* (*Lumbricidae*) occurs in damp soils along bodies of water, in bogs and fens and in stream beds. Thus it is possible that animals technically considered earthworms will be found as benthos.

Another habitat commonly used by certain species of earthworms in forests is decomposing logs. The animals may be found beneath bark or in pockets of organic soil formed in hollows of logs. Often the species involved are specialists on this habitat type, and will only rarely occur in the soil proper.

In ecosystems humid enough to support epiphyte growth on trees, or mats of vegetation on rocks, organic soils build up along tree limbs or on top of boulders which may be inhabited by earthworms. Tank epiphytes, such as many bromeliads, support earthworm species unique to this specialized location. Other plants, including palms, may accumulate deposits of organic debris in leaf axils. If the plants and deposits are sufficiently large and long-lived, they may support earthworms as well. Such microhabitats may not be soils in the strictest sense, but they function in the same manner, with plants rooting in them and soil organisms living in them. It is debatable whether the aquatic environment of a bromeliad tank does not qualify the debris within as a sediment. If a complete inventory of the soil organisms in an area is desired, then it would be best to include these above-ground 'soils' as well.

Recovery of earthworms from such locations, including logs, is best accomplished by manually searching the microhabitat in the field. Transport to laboratories will be impractical in most cases. If other smaller organisms are to be collected as well, then small subsamples of the microhabitat should be taken (e.g. tank epiphyte contents, rotten log material) for extraction in the lab.

Collection and Preservation of Earthworms

Collectors should be alert for evidence of earthworm presence, such as worm faecal material (casts) on the soil surface and within the soil, and earthworm burrows in the soil. It may take some practice to distinguish these from tunnels of termites and ants. Alertness to these clues will maximize success. Do not stay in one spot if nothing is turning up. Dig up blocks of soil at least 25-40 cm across, cutting around the edges as quickly and as deeply as possible. Depth of sampling is generally not great (10-30 cm) unless there is some sign of earthworms burrowing deeper or escaping from the shallower sampling. Soils from ridges and valleys should be sampled within the same general habitat type. Arboreal habitats include the tanks of bromeliads, rootmats formed by epiphytes, thick moss mats on tree limbs, boulders, etc. There may be vertical within-plant habitat separation in bromeliads, corresponding to different ages of leaves. Rotting logs with the bark still on should be inspected by taking off the bark. It is possible but not common to find worms in soft rotten wood.

Specimens should be collected separately by the microhabitat in which they were found (e.g. soil, bromeliad, log, etc.) and this information should be kept with the specimens. Up to twenty individuals per collection site per date per ‘eyeball’ morphospecies could be collected until there is some idea of the level of within-species variability.

The simplest procedure is to kill the worms in alcohol, either ethanol or isopropanol, at 50% (v/v) concentration and then immediately fix them in 10% (v/v) formaldehyde. Most worms will die within seconds, but care should be taken to place them in the alcohol individually. This is because many species will exude mucus and may bind themselves into a dense mass that will have to be disentangled prior to fixation. There should be about three times the volume of formaldehyde as worm bodies. The best containers are glass vials with polyethylene cap liners (leak-proof), heavy-duty 'Zip-loc' bags, or 'Whirl-pak' bags. The 'Zip-loc' bags are very useful if one gets a lot of worms or some really big ones. When using 'Zip-locs', have a covered plastic box large enough to hold the bags flat. This helps keep the worms somewhat straight during fixation. After they have spent 48 h in formaldehyde it is optional to transfer them to 70% (v/v) alcohol for long-term storage or transport. If the transfer to alcohol is chosen, notes should be taken on the colours of the earthworms prior to replacing the fluid, since some species rapidly lose their original pigmentation in alcohol. Though many earthworms will not lose colour in alcohol after formaldehyde fixation, some do. Virtually all earthworms will lose colour completely if not fixed in formalin prior to immersion in alcohol for long periods. Alcohol preservation without fixation in formaldehyde generally results in soft, macerated and scientifically useless specimens.

Preservation in the field guards against the worms or parts thereof decomposing during long hot vehicle rides. After collecting has finished in a given location, the worms should be preserved and notes taken on their appearance, colour and behaviour before these details fade from memory. Photographs would be best taken of live worms or shortly after killing.

The only time formaldehyde should be avoided is where material is being preserved for biochemical or genetic analysis, such as DNA recovery and sequencing. In this case, alcohol should be at least 95% concentration and should be replaced once within 24 h. The specimens should be very stiff from dehydration caused by the strong alcohol.

Formaldehyde should not be stored long-term in plastic, since it seems to lose strength, but it is fine to take it to the field in plastic bottles. Do not transport it in squeeze bottles or anything that will leak. Use rubber gloves when handling formaldehyde or anything containing formaldehyde. Assume that it will leak and take the necessary precautions. Finally, work in a well-ventilated location.

An alternative to the above preservation procedure is to substitute a period of narcotization or anaesthesia for killing in alcohol. This produces specimens that are extended rather than contracted, and should be adopted if extreme contraction is observed when the worms are killed in 50% (v/v) alcohol. Narcotization can be done with dilute alcohol, perhaps 5% (v/v) at first (batch processing is allowed in this case), but sufficiently weak that the worms do not react violently and then die. They may be irritated and then calm down. Then, while gently stirring the dilute alcohol and worms, add a small quantity of alcohol and wait until any reaction has subsided. Continue adding small amounts of alcohol until the worms are totally unresponsive to stimulation. They can then be finished off by fixing them in formalin. The advantage of producing extended specimens is that taxonomic work is easier on extended specimens. However, most worms do not contract so severely that narcotization is necessary. An alternative to alcohol for narcotization is to use chloroform. The worms should be placed in a covered container in which is placed a ball of cotton wool with a few drops of chloroform on it. The cotton wool should be suspended such that the worms will not physically contact it.

Minimum Equipment and Supplies for Earthworm Collecting and Preservation

- spade
- plastic box for use during collecting
- camera(s) with flash and macrolens
- slide film
- formaldehyde (37% v/v)
- alcohol
- water

- storage bottles for solutions
- pan, bowl or other container for killing or narcotizing earthworms
- 'Zip-loc' bags, 'Whirl-paks' and/or vials with polyseal liners
- alcohol-proof, waterproof label material
- pencils
- rubber gloves
- forceps

Quantitative Sampling of Earthworms

In the event that estimates of earthworm species populations are required, it is most appropriate to use digging and sorting by hand or a combination of digging and a behavioural extraction. Population estimation techniques are reviewed in Bouche and Gardener (1984) and Springett (1981). Estimates are based on a standard sample unit of soil (25 × 25 cm to a depth of 30 cm) replicated at 5 m intervals along a transect or on locations randomly selected from a grid laid out on the site. If deep-dwelling species are known or suspected to live on the site and are not recovered by digging, then a chemical extractant can be poured into the hole. However, care should be taken not to disturb potential burrow channels at the bottom of the hole. The edges of the soil sample should be cut as deeply as possible as quickly as possible all around the perimeter of the sample. Doing so will disrupt the horizontal burrows and decrease the escape routes of the animals. Cutting around the edges also severs roots and simplifies removal of the soil. The soil sample can be placed in buckets, trays, or on sheets of plastic on the ground.

The soil should be searched carefully and, if time allows, searched twice. Small amounts of soil should be searched at a time and then transferred to a separate container or location. It may be desirable to take soil in 10 cm layers, but this will require very careful excavation of the soil sample. In forests this may be extremely difficult on account of the roots of trees. Generally the majority of earthworms will be in the upper 10-15 cm provided the soil is moist. Greater searching efficiency will be had if the upper layer of soil is not mixed with the lower layer and is searched separately, whether or not one intends to collect and record earthworms by depth.

Since earthworm numbers per sample unit tend to be randomly distributed, at least 10 sample units per site should be taken, and the data transformed appropriately before analysis. Unitary habitats like bromeliads could be measured for size and worms collected and stored plant-wise. Quantitative sampling will require more time to achieve the same degree of habitat/ecosystem coverage. If quantitative sampling is considered desirable, it should be done after the species and their periods of reproductive maturity are known. That would necessitate visiting sites again after making an initial qualitative survey and identifying the species present.

Identification

If qualitative sampling is done, then more effort should be expended in an attempt to identify the individuals to species, because otherwise the data will amount to a declaration that earthworms are present in the system. Identification is not technically difficult, but is badly hindered by the lack of comprehensive, up-to-date keys to the earthworm genera and species of the world. The most recent such key is that of Michaelsen (1900, in German), though Gates (1972) allows the reader to reach the families as proposed by that author. Unfortunately, this situation is not likely to change in the near future. Michaelsen (1900) does allow identification fairly close to the genus, and is the starting-point for any totally unknown specimen. As with many other invertebrate taxa, specialists on earthworms are few and far between. A training programme to create 'parataxonomists', after the model of the Instituto Nacional de Biodiversidad (INBio) in Costa Rica, would relieve much pressure on the specialists and accelerate the pace of specimen identification and species description. Ecological niche categories can be approximated by examination of simple morphological features (e.g. Lee, 1959; Bouche, 1977) and the specimens divided along those lines to give a more detailed breakdown of the earthworm community (Fragoso and Lavelle, 1992).

Limitations of Existing Techniques

Behavioural techniques share a number of disadvantages. When they work depends to some degree on the activity status of the worms and on the nature of the soil. Obviously, if worms are inactive due to temperature or drought, they will not emerge. In the case of chemicals, if the soil is very dense and penetration of the solution is slow, and if burrows are not open at the surface, worms will not be inspired to come out, or will do so only very slowly, or they may die in the soil before emerging. It may not be practical to work with solutions on steep slopes.

A potential problem with all chemical extraction methods is the effect on non-target organisms. They may be killed outright or the effect may interfere with other attempts to recover biota from the soil. Then there is the environmental and health hazard of using very toxic or carcinogenic chemicals in this manner, particularly if the same soil is to be manually searched after extraction in order to assay the effectiveness of the method.

The primary disadvantage of the formalin method is that it is not generally effective. Though it works very well on the *Lumbricidae*, it is not useful for all other earthworm families. It does not work at all on most *Megascolecidae* (*s.l.*), the largest earthworm family. Persons interested in formalin extraction should experiment with it before electing to use it routinely in the field. Care should be taken to calibrate its effectiveness on the various species present. The same can be said of all other chemicals used as extractants - that their taxonomic range is unknown and perhaps limited.

Electrical extraction is potentially hazardous to the operators. Faulty equipment sold for this purpose in the USA has killed at least one person intent on going fishing with earthworms for bait. Care should be taken to avoid electrical shocks.

Integrating Techniques into a General Scheme of Analysis

Earthworms and other soil macrofauna typically exist at low densities compared with microarthropods, nematodes, protozoa, bacteria and fungi. While the latter small organisms may be adequately sampled from small volumes of soil, the macrofauna cannot. In order to adequately sample earthworms within a programme intended to cover all soil and sediment biota, separate soil samples will have to be taken. It would be feasible, within the limits of extraction and culturing techniques, to sample microbiota from soil already searched for earthworms, but the reverse is not likely to bear fruit. A sampling protocol for earthworms should include field-searching several relatively large volumes of soil (see above for suggested sample unit sizes and numbers) in conjunction with one or more behavioural extraction techniques proven effective on the local earthworm fauna. It may be possible to integrate this with sampling for other macrobiota, depending on the methods favoured for the other taxa. Subsamples of the soils taken for the macrobiota can be used for all other soil/sediment biota.

Ecosystem, habitat and microhabitat coverage within a nation or region will no doubt be decided based on the resources and expertise available to support the sampling programme, and to some extent based on the expected among-habitat variation in species composition of the various soil taxa. Earthworms are relatively sedentary organisms with no long-distance dispersal of individuals or gametes. Consequently, a thorough sampling programme should include all ecosystem types and all isolated occurrences of those ecosystem types, plus distinguishable within-ecosystem habitats. For example, riparian areas and upland sites should be sampled within a forest type, and that forest type should be sampled separately wherever it is found geographically isolated from other areas of that forest type. If there are two mountains with ecosystem zonation on their slopes, and a lowland area between them, thoroughness would dictate sampling on both mountains. This may not be considered necessary for more vagile taxa.

Non-soil or other unconventional habitats for earthworms add another dimension of sampling. There needs to be some consideration of the limits of the sampling programme as regards these other homes of 'soil organisms'. The distinction between soil and not-soil is often arbitrary, and needs precise definition, although this may be difficult.

Bibliography

Andrews, W.A. (1972) *A Guide to the Study of Soil Ecology.* Prentice-Hall, Englewood Cliffs. [This provides basic methods for sampling soil organisms.]

Blair, J.M., Parmelee, R.W. and Lavelle, P. (1995) Influences of earthworms on biogeochemistry. In: Hendrix, P. (ed.) *Earthworm Ecology and Biogeography in North America.* CRC Press, Boca Raton, pp. 127-158. [An excellent and current review of the effects of earthworms on macronutrient movements, carbon flux in and out of soils, and other topics related to soil nutrient status.]

Bohlen, P.J., Parmelee, R.W., Blair, J.M., Edwards, C.A. and Stinner, B.R. (1995) Efficacy of methods for manipulating earthworm populations in large-scale field experiments in agroecosystems. *Soil Biology and Biochemistry* (in press). [Details of the electrical shock technique.]

Bouche, M.B. (1977) Strategies lombriciennes. In: Lohm, U. and Persson, T. (eds) *Soil Organisms as Components of Ecosystems. Ecological Bulletin (Stockholm)* 25, 122-132. [Bouche's important work describing the three basic ecological categories of earthworms and how to judge them from morphological and behavioral characteristics.]

Bouche, M.B. and Gardener, R. (1984) Earthworm functions VII. Population estimation techniques. *Revue d'Ecologie et de Biologie du Sol* 21, 37-63. [This is a useful review and description of methods to estimate earthworm populations.]

Daniel, O. (1992) Population dynamics of *Lumbricus terrestris* L. (*Oligochaeta: Lumbricidae*) in a meadow. *Soil Biology and Biochemistry* 24, 1425-1431. [The author provides a specific example of application of the formalin and chloroacetophenone extraction methods on a deep-burrowing species.]

Eisen, G. (1900) Researches in American Oligochaeta, with especial reference to those of the Pacific Coast and adjacent islands. *Proceedings of the California Academy of Sciences,* Third Series II, 85-276. [This largely taxonomic work includes notes on collection techniques and a case of human use of earthworms for food.]

Fragoso, C. and Lavelle, P. (1992) Earthworm communities of tropical rain forests. *Soil Biology and Biochemistry* 24, 1397-1408. [The paper is a good example of how one might go about describing the ecology of the earthworms of forest site. It is relevant to this manual because the authors began in a system totally unknown for earthworms.]

Fragoso, C., James, S.W. and Borges, S. (1995) Native earthworms of the north neotropical region: current status and controversies. In: Hendrix, P. (ed.) *Earthworm Ecology and Biogeography in North America.* CRC Press, Boca Raton, pp. 67-116. [This reviews north neotropical earthworm biogeography and outlines some of the areas critically in need of further research.]

Gates, G.E. (1972) Burmese earthworms. *Transactions of the American Philosophical Society* 62, 1-326. [Primarily, a taxonomic work, but with valuable sections on methods of study of earthworm morphology, some of the major concerns in systematics, and an excellent glossary of terms used in systematic literature on earthworms.]

James, S.W. (1991) Soil, nitrogen, phosphorus and organic matter processing by earthworms in tallgrass prairie. *Ecology* 72, 2101-2109. [This paper provides a field and experimental approach to earthworm contributions to ecosystem processes in a natural ecosystem.]

Kempson, D., Lloyd, M. and Ghelardi, R. (1963) A new extractor for woodland litter. *Pedobiologia* 3, 1-21. [The basic Berlese funnel method in a slightly modified form is available here.]

Lavelle, P and Martin, A. (1992) Small-scale and large-scale effects of endogeic earthworms on soil organic matter dynamics in the humid tropics. *Soil Biology and Biochemistry* 24, 1491-1498. [A very important paper as regards an understanding of how earthworms influence soils organic matter. Organic matter protection takes place after an initial phase of decomposition.]

Lee, K.E. (1959) The earthworm fauna of New Zealand. *New Zealand Department of Science and Industrial Research Bulletin* No. 130. [Lee described three ecological types of earthworms well before Bouche, but his contribution was largely overlooked until Bouche's (1977) paper. Nevertheless, it shows the generality of the phenomenon and comparing the two papers one learns the limitations of the categorization, especially when comparing across families.]

Lee, K.E. (1985) *Earthworms: Their Ecology and Relationships to Soils and Land Use.* Academic Press, New York. [This book is the best currently available comprehensive review of earthworm ecology.]

Michaelsen, W. (1900) *Oligochaeta.* In: *Das Tierreich.* Lief. 10. R. Friedlander und Sohn, Berlin. [A taxonomic volume still indispensable for identifying earthworms, particularly from poorly known areas such as most of the tropics.]

Moore, A.M. and Burns, L. (1970) Appendix C: Preliminary observations on the earthworm populations of the forest soils of El Verde. In: Odum, H.T. and Pigeon, R.F. (eds) *A Tropical Rain Forest.* NTIS, Virginia, p. 238. [Though sketchy in details, and lacking any identification of the species involved, the remarkable amount of earthworm biomass found in the Puerto Rican forest suggests the potential importance of earthworms to tropical forest ecology.]

Raw, F. (1959) Estimating earthworm populations by using formalin. *Nature* 184, 1661-1662.

Smith, F. and Green, B.R. (1916) The Porifera, Oligochaeta, and certain other groups of invertebrates in the vicinity of Douglas Lake, Michigan. *Michigan Academy of Sciences Annual Report* 17, 81-84. [Earthworms don't only live in the soil; they may be found in sediments as well.]

Springett, J.A. (1981) A new method for extracting earthworms from soil cores, with a comparison of four commonly used methods for estimating earthworm populations. *Pedobiologia* 21, 217-222.

Swift, M.J. and Anderson, J.M. (1989) Decomposition. In: Lieth, H. and Werger, M.J.A. (eds) *Tropical Rain Forest Ecosystems. Ecosystems of the World-14B.* Elsevier, The Netherlands, pp. 547-569. [In tropical systems, earthworms and termites are the most important macrofauna involved in decomposition of plant residues. Data for earthworms are much less than for termites.]

Marine Macrofauna: Polychaetes, Molluscs and Crustaceans

21

ANDREW S. Y. MACKIE and P. GRAHAM OLIVER

Department of Zoology, National Museum of Wales, Cathays Park, Cardiff CF1 3NP, UK.

The Importance of the Macrofauna in Marine Ecosystems

Polychaete worms, molluscs and crustaceans are the dominant macrofaunal groups in all marine sediments from the intertidal to the deep-sea. They are vital to the structure, production, dynamics and health of the marine benthos and the wider marine environment. They aid the deposition, breakdown, incorporation and turnover of organic matter in the seabed, helping to recycle nutrients to the overlying water column. The majority of species are small and short-lived, exhibiting a high secondary production. Hence they are an important link in marine food webs and feature greatly in the diets of larger, more mobile, predators, some of which (e.g. demersal fish such as plaice) have commercial importance in fisheries. Larger macrofaunal bivalve and gastropod molluscs, shrimps and crabs, and a few polychaetes are also harvested directly. On European intertidal estuarine mudflats enormous numbers of relatively few species, e.g. ragworm (*Nereis diversicolor*), snail (*Hydrobia* spp.), the Baltic clam (*Macoma balthica*) and the 'mud-shrimp' (*Corophium* spp.), support large populations of birds. In addition, it is likely most of the planktonic eggs and larvae of many species are eaten by pelagic animals. Some species are collected or cultured as bait for fishermen, e.g. the king ragworm, *Nereis virens*. Polychaetes have also been used to remove organic wastes from aquaculture systems (Tenore *et al.*, 1974) and as toxicological test organisms (Reish, 1980).

Apart from the ecological and commercial aspects, the benthic macrofauna can be utilized to monitor the effects of pollution. The majority of species are essentially sedentary and therefore changes in their community structure and diversity can be examined in relation to inputs of pollutants (Warwick, 1993;

Warwick and Clarke, 1993). Many contaminants are known to accumulate in seabed sediments, particularly those with higher mud contents. Further advantages of utilizing the macrofauna are that the literature available is usually comprehensive enough to allow identification to at least genus level, and that the animals themselves are easily examined using simple light microscopes.

Size Categories of Marine Invertebrates

In the marine environment the sublittoral soft-bottom habitat is dominated by invertebrate animals. For practical purposes these are conveniently subdivided into three broad size classes: meiofauna, macrofauna and megafauna. The separation of these groups is not well defined, though there is some evidence for size partitioning between them (Schwinghamer, 1981; Warwick, 1984). In reality the size distributions of the categories overlap.

The meiofauna, with the smallest body size, is generally considered to include animals passing through a 0.5 mm mesh sieve. Depending upon the meiofaunal taxon under investigation sieve meshes down to 30-40 μm may be necessary (McIntyre, 1969; McIntyre and Warwick, 1984). The macrofauna is often considered to include those taxa retained on a 0.5 mm sieve, though in some situations larger or smaller meshes may be deemed necessary (see below). The megafauna consists of very large animals that can readily be sampled by hand.

The majority of the benthic macrofauna encountered belongs to four major taxonomic groups: *Polychaeta* (bristleworms), *Mollusca* (snails and bivalve shells), *Crustacea* (crabs, shrimps, sandhoppers, etc.) and *Echinodermata* (starfish, sea urchins, sea cucumbers). The relative abundances of these varies from locality to locality and faunal assemblages may be numerically dominated by any of the groups. Commonly, however, the polychaetes will be the most prominent enumerable taxon, with the molluscs or crustaceans ranked second. For example, in a recent study in the Irish Sea (Mackie *et al.*, 1995) the overall rankings were: *Polychaeta* (53% individuals, 49% species), *Mollusca* (27% individuals, 19% species), *Crustacea* (8% individuals, 22% species), all other invertebrates (12% individuals, 10% species).

Sampling/Extraction Techniques

This section primarily concerns the collection, field processing, fixation, preservation and laboratory processing of sublittoral benthic macrofaunal samples, with special emphasis on the polychaetes, molluscs and crustaceans. Most of the procedures are equally applicable (sometimes with modification) in other marine environments (e.g. the intertidal, deep-sea). A number of comprehensive reviews are available concerning sampling apparatus, sampling design and sampling processing. Key works are Elliott (1977), Green (1979), Holme and McIntyre (1984) and Baker and Wolff (1987). Detailed descriptions of particular aspects can be obtained by reference to the Bibliography. Emphasis here has been placed upon making recommendations derived from our own practical experience.

Sampling Strategy

The design of any sampling programme is determined by the questions its execution is required to answer. Useful reviews and information can be found in Baker and Wolff (1987), Bamber (1995), Burd *et al.* (1990), Cuff and Coleman (1979), Elliott (1977), Green (1978, 1979), Hartley (1989), Höisaeter and Matthiesen (1979), Holme and McIntyre (1984), Kingston and Riddle (1989), Lie (1968), Rees *et al.* (1990, 1991, 1994), Riddle (1984, 1989a), Scherba and Gallucci (1976), Smith (1978) and Wildish (1978).

Some of the basic questions that must be asked before commencing a survey include:

- What is already known about the proposed area of study? Much time and money can be saved by thorough preparation. If nothing is known, it may be desirable to carry out a simple low-cost preliminary programme prior to the main onc.
- Are quantitative or qualitative data required? If quantitative, then what degree of statistical precision is necessary? This will influence the number of replicates and size of samples required per station.
- Are the samples and/or stations to be arranged spatially in a pattern that will enable the requirements of the study to be met? For example, stations could be arranged in a transect, regular grid, stratified pattern or simply randomly.
- Are control stations required? If the purpose of the study is to monitor the effect of a source of pollution this would clearly be desirable.
- To what extent is the project influenced by temporal change? Population densities may fluctuate greatly throughout the year and between years due to natural processes (e.g. success of reproduction, larval settlement, weather conditions).
- Are the equipment, personnel, expertise (particularly taxonomic), time and funding available to successfully complete the study?

The basic protocol is to take samples of the sediment, usually quantitatively, and aim to retain all the fauna within each. The animals are then extracted from the sediment and preserved. Qualitative sampling is often advantageous as an initial estimate of species richness. Generally some dredge or trawl device is used and this may partially separate the fauna from the sediment as it collects the sample.

Collection Methodology

Good accounts of polychaete specimen collection and treatment can be found in Fauchald (1977) and George and Hartmann-Schröder (1985); those in the latter are repeated in Pleijel and Dales (1991). Westheide (1990) described suitable procedures for dealing with the smaller meiofaunal polychaetes. Eleftheriou and Holme (1984), Giere and Pfannkuche (1982), McIntyre and Warwick (1984) and Hartley *et al.* (1987) provide valuable descriptions of whole fauna techniques. The reader is referred to these for additional information and references concerning many of the points discussed below.

All aspects of macrofaunal sampling are governed by the aims of the study, its cost, the time available and the equipment, personnel and expertise available.

Samplers

There have been a number of extensive comparative studies concerning sampling devices and their efficiencies. Key works are Ankar (1977), Beukema (1974), Birkett (1958), Eleftheriou and Holme (1984), Elliott and Tullett (1978), Gage (1975), Gallardo (1965), Kutty and Desai (1968), Lie and Pamatmat (1965), Riddle (1984, 1989b), Rosenberg (1978), Rumohr (1990), Tyler and Shackley (1978), Wigley (1967) and Word (1976).

To efficiently sample the benthic macrofauna a sampler must be able to penetrate to a depth sufficient to capture the animals present. Numerous studies of the vertical distribution of animals within the sediment have been carried out (e.g. Sanders, 1960; Lie and Pamatmat, 1965; Jumars, 1978; Shirayama and Horikoshi, 1982). Most report the majority of species and individuals to be present in the upper 5 or 10 cm, although large burrowing molluscs and crustaceans may be found deeper. These large species are generally undersampled using the most commonly used techniques and, in studies involving biomass estimation, additional specific sampling methods may be required. A few workers (Johnson, 1967; Christie, 1976a,b; Hines and Comtois, 1985) have recorded some small species deep within the sediment.

An ideal sampler would, therefore, be some form of coring device. The best available is the 0.06 m^2 Reineck box sampler or the larger 0.25 m^2 deep-sea spade corer (e.g. see Eleftheriou and Holme, 1984; Rumohr, 1990). One problem with such apparatus is the need for relatively large ships for their operation. Consequently, most inshore workers rely on grab samplers, which typically take samples of 0.1 m^2 area, sometimes larger or smaller depending upon their design. Of these, the most efficient is the long-arm continuous-warp Van Veen grab (see Riddle, 1989b; Rumohr, 1990), which benefits from a good initial penetration. Small hand cores (5-10 cm diam.) can be used on intertidal sediments and by divers in shallow waters. Soft cohesive sublittoral sediments can be remotely sampled by multiple corers (Barnett *et al.*, 1984; McIntyre and

Warwick, 1984), which can take up to twelve 5-7 cm diam. cores at each deployment.

Qualitative samplers include trawls, dredges and traps. Each is selective in relation to the size, type and behaviour of the fauna sought. For most samplers the depth of sediment penetrated will have a major influence on the animals captured.

As a general tool for sublittoral surveys a 1 m wide rectangular framed trawl with a 1 cm mesh bag is recommended for taking large surface and shallow-burrowing fauna. In deep-sea research much larger versions of the Agassiz trawl (up to 3 m wide) are still used to sample the scarce megafauna.

For taking the deeper infaunal animals a digging dredge is more efficient, especially in sediments where grab penetration is poor, e.g. hard-packed sands, gravels and stony grounds. A double-edged rectangular anchor dredge design, with an inner 0.5 mm mesh bag, is recommended if it is to substitute for a grab. A larger 1 cm mesh is adequate if only the larger fauna is sought. The dimensions of the dredge are entirely dependent on the size of ship available to deploy it. In deep-sea research, sophisticated versions of the dredge have been developed where epibenthic sledges with automatic closing and 300 µm mesh bags collect the fauna from the surface and top 5 cm of sediment.

Size, number and volume of samples

The size and number of samples taken depends largely upon the aims of the study. For example, in a general sublittoral grab-sampling survey designed to map the invertebrate communities of an area of seabed, a single 0.1 m^2 sample per station may be adequate (Cuff and Coleman, 1979; but see Green, 1980 and Cuff, 1980), though most workers take two or three replicates (e.g. Mackie *et al.*, 1985). Alternatively, an investigation concerning temporal change or quantitative comparison will require perhaps five to ten such samples per station (Lie, 1968; McIntyre *et al.*, 1984; Riddle, 1984, 1989a) in order to meet an acceptable level of statistical precision. Kingston and Riddle (1989) gave examples of situations where diversity index estimates from single sample stations satisfactorily revealed pollution gradients.

Optimization of sample size and number has been subject to much theoretical debate and some practical investigation (e.g. Lie, 1968; Westerberg, 1978; Downing, 1979; Riddle, 1984, 1989a; Bros and Cowell, 1987; Vézina, 1988). It is recognized that, for a smaller total area sampled, many small replicates can produce the same statistical precision as obtained using a lower number of large samples. Downing (1979; see also 1980a) calculated potential reductions of 1333-5000% in the total area sampled by using a 20 cm^2 sampler rather than a 0.1 m^2 one.

Downing's paper generated much discussion (Taylor, W.D., 1980; Downing, 1980b, 1981; Taylor, L.R. 1980; Chang and Winnell, 1981; Vézina, 1988) and it has been found that his estimates may not hold in all situations. Riddle (1989a), working on a low-density sand community, found that 226 cores of 4.75 cm diam (total area 0.4 m^2) would be needed to achieve the same standard of precision (standard error 20% of the mean abundances for 30% of the species) as five 0.1 m^2 grab samples. The saving in this case was therefore only 20%. Hence, for such a strategy to be viable, the reduction in sieving, sorting and identification time has to outweigh the effort required in taking the more numerous samples.

As most infaunal invertebrates are to be found in the upper 5-10 cm of sediment a sampler should preferably sample to 10-15 cm depth. Riddle (1989b) found that, on medium to fine sand, a heavy long-armed Van Veen grab collected to a relatively uniform depth of about 10 cm over the 0.1 m^2 area sampled. Thus each replicate would collect about 10 litres of sediment. Other grab samplers take different-shaped bites and the volumes taken relate differently to the depth sampled. Corers sample to uniform depths across their area; therefore, in the example of Riddle (1989a), 226 cores penetrating to 10 cm would collect a total of 40 litres of sediment.

The depth of sediment collected by samplers varies according to sampler and the nature of the substrate. Consequently, benthic workers have to decide the minimum 'acceptable' for their locality and the aims of the investigation. In North European studies of areas with 'hard' substrates this may be set as low as 4-5 cm, equivalent to 4-5 litres of sediment for a Van Veen grab (e.g. Riddle, 1984; Hartley and Dicks, 1987; Mackie *et al.*, 1995). On very hard stony ground it may be difficult to get more than 3 cm penetration. On the softest muds the grab may collect some animals at a maximum 15-16 cm depth; however, because of its shape and bite profile, the total volume sampled cannot exceed about 11-12 litres. Box corers can efficiently sample to 40 cm in muddy sediments (Hessler and Jumars, 1974), but penetrate less on sand and do not function well on harder grounds.

Sieve size and sieving

The size of sieve mesh used is effectively a compromise between cost and accuracy. In early studies much work concentrated on the larger conspicuous animals. Thus, benthic ecologists using meshes of 2 mm or more largely ignored the majority of the species now referred to as macrofauna.

There have been a number of investigations into the capture efficiency of different sieve mesh sizes (e.g. Reish, 1959; Rees, 1984; Bishop and Hartley, 1986; Bachelet, 1990). It is clear from these studies that 'macrofauna' cannot be defined by mesh size, despite the findings of Schwinghamer (1981) and Warwick (1984). Detailed discussion of the problem can be found in Bishop and Hartley (1986), Hartley *et al.* (1987) and Bachelet (1990). The former acknowledge the inefficiency of 1 mm mesh sieves, but regarded it as the most cost-effective size to use in studies involving faunal abundance and diversity (see also Kingston and Riddle, 1989). This has been largely followed by monitoring groups, though in time-series investigations a 0.5 mm mesh has been recommended in order to more accurately assess population dynamics (Rees *et al.*, 1990). In his important work, Bachelet (1990) found differences in the relative capture efficiencies between phyla in the ranking crustaceans > polychaetes > molluscs. He also noted that a 1 mm mesh was suitable only for biomass estimates and that, for 'complete' population dynamic studies, 100-200 µm meshes would be required. It is of interest here that the sieve meshes routinely used in deep-sea studies are 250 or 300 µm (e.g. Grassle and Maciolek, 1992; Blake and Hilbig, 1994; Paterson *et al.*, 1994). Perhaps all workers should standardize on one size, possibly 300 µm? However, this is debatable.

Bachelet (1990) acknowledged the practical problems of using very fine sieves and suggested supplementing 1 mm sieving with extra small samples or subsamples being examined using a smaller mesh. His third alternative of using flotation (i.e. elutriation) may be more profitable. Mackie (1994) and Mackie *et al.* (1995) employed a sample fractionation technique based upon elutriation, the smaller animals being passed into the sieve prior to the bulk of the sediment. This removed the vast majority of the fauna and left relatively few animals in the residue. It may therefore be that a fine sieve (e.g. 0.3 or 0.5 mm) could be used for the initial fractionation and the remaining sediment screened by a coarser one (e.g. 0.5 or 1.0 mm). Potential problems with such a scheme include ensuring adequate initial elutriation and, in some sandy situations, a fauna adapted to disturbance-avoiding capture by adhesion to sediment particles.

Sieving operations will be made most efficient if the screening area is as large as manageable (e.g. 45 cm diam.). This is especially so for sieves with mesh sizes less than 2 mm. There should be enough sieves available (minimum of two) so that clogging of, or damage to, one sieve does not delay the sieving procedure.

Faunal extraction

Once the mesh size has been set, the sediment is washed through the sieve and the fauna and certain sediment particles retained. Two factors arise from this process. First, sieving can damage the animals (especially delicate polychaetes and crustaceans) and render identification very difficult, if not impossible. Second, the smaller the mesh used the more sediment retained and this extends the sorting time. To avoid these problems, various methods of faunal extraction have been developed utilizing elutriation, differential density (e.g. using sugar solutions, 'Ludox', carbon tetrachloride) or the avoidance reactions of the animals themselves (using seawater ice). All these techniques are described in detail in Eleftheriou and Holme (1984), McIntyre and Warwick (1984) and Giere and Pfannkuche (1982). Only the first is readily employed in the field and the success rate of the others is variable, with the seawater-ice method the least reliable, although Westheide (1990) recommended it as a means of obtaining interstitial polychaetes in good condition.

The differential density methods are generally only applicable for separating specimens from preserved samples (see McIntyre and Warwick, 1984). 'Ludox', a proprietary colloidal silica polymer, has largely superseded the use of the carcinogenic CCl_4, though it is itself toxic and should only be used with care.

Elutriation involves suspending the sediment in copious quantities of water and decanting off the lighter, more fragile animals (mostly polychaetes and crustaceans). The more vigorous direct sieving is then used to extract the larger or heavier animals (mostly molluscs and echinoderms). It is essential that the water used is of equal salinity to that of the environment the animals were taken from. The use of fresh water on living marine animals can be disastrous, causing rapid fragmentation. At this stage it is also important to maintain the original temperature of the sediment-overlying water; marine invertebrates are often sensitive to increases in temperature. Thermal tolerance is generally greatest among animals from shallow tropical regions and from intertidal environments. Conversely, animals from colder waters with relatively low thermal ranges may die and rapidly decompose on exposure to elevated temperatures. Extraction of small interstitial invertebrates from small samples of sandy sediments can also be aided by pre-treatment with magnesium chloride, as a relaxant (see below).

Workers must always be aware of the purpose of the sieving procedure. **The important things are the animals**. Unfortunately this is often forgotten and the processing of samples focuses on getting the material through the sieve as fast as possible.

Marine invertebrates are in the main **very fragile creatures** which can easily fragment and various appendages (e.g. branchiae, antennae, palps, legs) fall off. This should be remembered at all times. Sieving can be very harsh. Blasting a sample through a sieve or grinding the animals about among lots of sand or gravel will clearly result in disintegration. **Invertebrates are easiest and quickest to identify when they are in good condition**. Very badly mangled specimens can be indeterminable. A little extra effort in the field will produce substantial benefits later in the laboratory. Naturally this has to be weighed against ship time and cost, but the quality of the work will certainly be better if more care can be taken.

Sample Treatment

Once the samples have been sieved, the integrity of the specimens must be maintained as far as possible. If the animals are required for live sorting then they must be covered with sufficient suitable water to ensure their survival. As mentioned above, temperature and salinity should ideally be kept near that of the collection site. If necessary an air pump may be used to oxygenate the samples. In certain cases, narcotization prior to fixation may be necessary to ensure intact specimens.

Narcotization

The use of specific relaxants can greatly improve the condition of the specimens obtained. A considerable number of narcotic agents have been employed in different invertebrate groups (see Lincoln and Sheals, 1979; Smaldon and Lee, 1979). Many are only used for specific taxa and, usually, the best results are achieved by application to individuals or small numbers of selected species. In this way, the relaxation process can be closely observed and the optimum time for fixation determined. Over-relaxation can be worse than none at all. Bulk-sample relaxation should be used only with great care, since there will be a differential effect on different animals; some will be over-relaxed, some not relaxed enough.

For the polychaetes (see Mackie, 1994) and many other invertebrates, magnesium chloride and menthol appear to be of most use. In addition, their low toxicity (if any), ready availability and low cost give them a great advantage over some other potentially dangerous substances, e.g. MS222 ('Tricaine'; ethyl *m*-aminobenzoate methane sulphonate). The effectiveness of each varies between different families, and sometimes between different species or genera within a family. Which to use is largely a case of trial and error.

Magnesium chloride (used as a 7% (w/v) solution in fresh water) can be gradually added to live specimens in seawater (up to about 50:50 $MgCl_2$ to seawater) until they stop moving. This relaxant is of good general use for the majority of polychaetes and molluscs. It is particularly useful for phyllodocids, syllids, hesionids, some spionids, poecilochaetids, sabellids and most other small fragile polychaetes. Note that certain species (e.g. some phyllodocids and terebellids) may exude much mucus in response to $MgCl_2$. However, with care, this can usually be teased away.

Menthol can be used by adding crystals to the surface of the liquid in the specimen dish or container and allowing them to slowly dissolve in the seawater. Finely ground crystals work more effectively than large crystals. Relaxation is also quicker when the container is covered. Alternatively, and with more rapid effect, menthol crystals can be mixed with seawater and allowed to stand for some time prior to use. The solution obtained from this can be used directly (up to about 50:50 menthol solution to seawater) and the undissolved crystals re-used when needed. Menthol works very well on polychaetes (scaleworms, hesionids, nereids, nephtyids, eunicids, terebellids and sabellids), gastropod molluscs and echinoderms.

Other more commonly used narcotics include ethanol (for polychaetes, molluscs and echinoderms) and propylene phenoxetol (for small polychaetes and crustaceans). Ethanol must be added gradually, drop by drop, to prevent violent avoidance reactions from the animals. These two relaxants are toxic and should be carefully handled. Smaldon (1978) has successfully used carbonated seawater to relax a wide range of invertebrates, including anthozoans, gastropods, bivalves and decapod crustaceans.

Fixation

The need to fix animals well is as important as treating them carefully. Inadequately fixed specimens can be almost impossible to identify, even by an expert. They will be very soft and pliable and in extreme cases will disintegrate while being manipulated by forceps under the microscope.

Often workers consult literature and find that 4% (v/v) formaldehyde in seawater (1 part 40% (v/v) formaldehyde (formalin) to 9 parts seawater; i.e. 10% (v/v) formalin) is recommended for fixing various invertebrates. Sometimes 2% (v/v) formaldehyde (5% (v/v) formalin) may be deemed sufficient (e.g. Ockelmann, 1989). This is satisfactory for animals placed directly in appropriate volumes of fixative solution. Note, however, that the volume and strength of the fixative must be carefully considered since water retained in the sieved sediment, in polychaete tubes, etc., and in the body fluids of the animals will dilute the formalin further.

There are two basic ways of approaching the problem of achieving a satisfactory concentration of fixative:

- Large volumes of 'standard' fixative solution can be added to the samples (e.g. 4 or 5 parts 10% (v/v) formalin to 1 part sample). Unfortunately, it is our experience that insufficient amounts of 10% (v/v) formalin in seawater are often used. This is especially so where large sieved sediment volumes and/or where tube-dwelling polychaetes are present. As a result the specimens are frequently poorly fixed.
- Smaller volumes of full-strength formalin are added to ensure penetration of enough fixative. For ease of use this method is commonly employed in offshore survey work (e.g. Mackie *et al.*, 1995). As before, the problem is calculating the 'correct' effective concentration of fixative in the formalin-seawater-sediment mixture. We would recommend that, to be sure, the estimated 'final' sample concentration be at least twice the minimum desired. Hence, 8-12% (v/v) formaldehyde (equivalent to 20-30% (v/v) formalin) should ensure good fixation. These concentrations may appear high but, quite simply, it is better to work on the safe side and add stronger, rather than weaker, fixative. The highest concentrations will be necessary for good fixation of tube- or crevice-dwelling invertebrates (including shell- or rock-boring species).

Whichever method of fixative addition is used, the sealed sample containers should be gently upturned/rotated to evenly distribute the formalin throughout the sieved sediment.

Formalin is acidic and animals left in it for too long may deteriorate, especially those having calcareous parts, which can dissolve away (e.g. molluscs and echinoderms). The rate at which deterioration occurs will vary depending upon the sieved sediment. For example, shelly sediments can help buffer the formalin. The simplest remedy is to thoroughly wash the samples in fresh water after at least a few days' fixing. This will remove the formalin and prevent any salt crystals developing on the animals. Buffers, such as borax (sodium tetraborate), can be added to the formalin before use, but sometimes precipitation causes problems. According to Harris (1992), no precipitation occurs when 40% (v/v) formaldehyde is buffered with 2% (w/v) sodium glycerophosphate.

For histological work, other fixatives, such as Bouin's fluid (71% saturated aqueous solution of picric acid:24% formalin:5% glacial acetic acid; available as a commercially prepared solution), will produce specimens in better condition. For scanning electron microscopy, glutaraldehyde in phosphate buffer is preferred, though formaldehyde can give acceptable results (e.g. Mackie and Chambers, 1990). Fixation in osmium tetroxide (2% (w/v) in seawater, for several hours) will give excellent results with polychaete specimens (e.g. Pleijel, 1990).

All these fixatives are toxic and great care must be taken in their use; gloves should always be worn. Osmium tetroxide fumes can fix eye and nose mucous membranes and should only be used in a suitable fume cupboard (Harris, 1992).

Staining and sorting

The stain rose bengal is commonly used in survey work. Although its use is depreciated by many pure taxonomists, it is certainly beneficial in aiding sorting and identification. Its use is virtually a necessity in large-scale survey work where contractual commitments necessitate the rapid production of quantitative data. Without staining, preserved invertebrates are generally white or yellowish and can easily be missed during sorting. Rose bengal is best added to the formalin before fixation: a strongly stained solution will produce bright red-coloured specimens.

Taxonomists may also need to stain their specimens for research purposes. The shapes of lamellae, presence of antennae, number of branchiae, etc. of white specimens are often difficult to determine accurately under the light of the dissecting microscope. The use of a temporary stain (e.g. methyl green) can resolve this problem. Such stains are taken up by the tissues, but leach out in alcohol. The leaching process can be prolonged by placing the specimen in reduced-strength alcohol or fresh water.

For rose bengal-stained specimens the initial sorting (to phylum level) can be done simply, using white trays and bench lights. However, this requires a good eye and often workers may prefer to use a dissecting microscope. In either case a key factor for success is not to place too much sample in the sorting tray/dish at any one time. The more material, the greater the chance of overlooking specimens. Sorted residues should be double-checked prior to disposal. To prevent damage to fragile-shelled molluscs, pliable stork-bill forceps must be used. The specimens must be covered by liquid (water or alcohol as appropriate) at all times as most invertebrates will completely distort if allowed to dry out, and subsequent identification may not be possible.

Secondary sorting (to family level) must be done under the dissecting microscope, preferably one with a zoom capability (e.g. 6-40 × magnification). Finally the invertebrates of a survey can be identified to species, family by family, using both dissecting and compound microscopes as necessary.

Preservation

Following the removal of the formalin (washed out with fresh water), the specimens or sievings must be placed in preservative to prevent decomposition and growth of bacteria or fungi.

The best preservative to use is 70-80% (v/v) alcohol (ethanol, isopropyl alcohol or industrial methylated spirit (IMS: 95% ethanol:5% methanol)). The volume of preservative should be several times that of the specimens or sample to avoid undue dilution. To ensure an adequate final concentration of preservative, the original alcohol can be replaced with new after a few days. Many specimens will retain their shape on direct transfer to 70% (v/v) alcohol; however, some delicate or poorly fixed invertebrates may distort as water is drawn from their bodies. This can be avoided by initially using weaker alcohol (e.g. 30% v/v) and, after an hour or so, replacing this with 50% (v/v), followed subsequently by 70% (v/v).

Propylene phenoxetol (1-2% (v/v) aqueous solution) has been used by many laboratories in recent years, but polychaetes do not seem to fare well under this treatment, being rather soft. On the other hand, alcohol can have the 'disadvantage' of making the animals too hard. This can be avoided by adding a little glycerol (1% v/v) or, preferably, propylene glycol (2% v/v) to the alcohol before use. Glycerol makes the specimens slippery and difficult to handle. Further, should the ethanol evaporate, the increasing glycerol concentration will clear the specimens and the final glycerol-water residue will encourage fungal growth (Levi, 1966). Propylene glycol protects specimens from completely drying out and inhibits fungi (Harris, 1978, 1992). Boase and Waller (1994) recommend topping up specimen vials containing ethanol-water-propylene glycol solutions with additive free alcohol. This will ensure that the propylene glycol concentrations do not progressively increase with time, although it was thought concentrations up to 5% (v/v) were routinely acceptable.

General Sampling and Extraction Protocol

The following is a synthesis of the protocol that we recommend for sublittoral and shallow marine infaunal benthos (Fig. 21.1).

Taking consideration of all the methods described, a sample processing scheme involving fractionation is recommended. The animals obtained will be

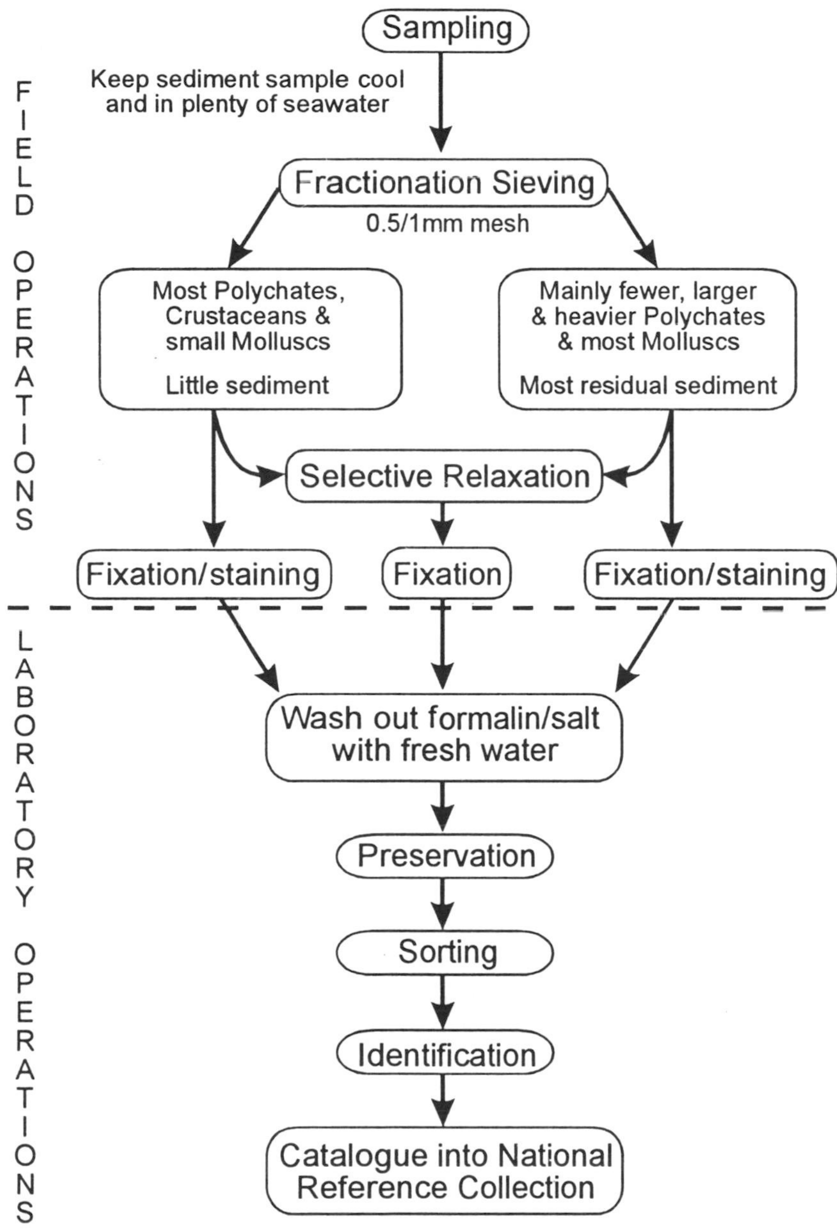

Fig. 21.1. Extraction protocol for sampling sublittoral and shallow marine infaunal benthos.

in better condition and identification consequently easier. The following scheme is detailed for a sublittoral sampling survey using grabs; however, the same principles hold for other situations (e.g. intertidal sampling; diver-collected cores). Much useful information concerning intertidal and diver sampling can be found respectively in Wolff (1987), and Gamble (1984) and Hiscock (1987).

Fractionation

Grab and dredge samples are emptied into large plastic fish boxes and immediately covered with seawater. This keeps the samples cool and initiates the gentle break-up of the sediment. Each sample is decanted into a large wooden hopper or high-sided tray and gently filled with copious amounts of seawater. Once the tray is half to two-thirds full, the water is released through the exit chute in a controlled manner and sieved using 45 cm diam. 0.5 mm mesh sieves. This procedure is repeated a number of times, gradually breaking up the sediment, until the majority of the mud and suspended specimens are removed. The material retained by these initial washings contains most of the delicate worms and crustaceans, as well as the smaller molluscs, but relatively little sediment. It is placed in a labelled container and fixed in a sample concentration of about 6-8% (v/v) formaldehyde (equivalent to 15-20% (v/v) formalin) in seawater.

The remaining unsieved sample fraction contains only the coarser sediment particles and the larger macrofauna, and can therefore be sieved with more vigour. This material can be further fractionated by placing a 2 mm mesh sieve above the finer one; each fraction being separately fixed in a sample concentration of about 12% (v/v) formaldehyde (equivalent to 30% (v/v) formalin). Once the fixative is added, each sealed sample container is gently upturned and rotated to distribute the formalin evenly throughout the sieved sediment.

Therefore, a sample from a sandy gravel station may be fixed as three separate fractions: washings, sand and gravel. Such fractionation greatly improves the quality of the sieved specimens and aids the later sorting phase. To help this process, most of the formalin can be strongly stained with rose bengal.

All containers (sample or sample fraction) should be identifiable by marking the outside with indelible pen and robust chemically stable labels (written in pencil or in Indian ink) must be placed inside each. On no account should only the container lids be labelled as, once removed, the samples will become unidentifiable.

At all stages of the sieving procedure noticeably fragile animals (e.g. scaleworms, phyllodocids, terebellids, nudibranchs) can be individually removed. These can be relaxed (menthol or magnesium chloride) before fixation.

Sorting and Identification

After at least 24 h (usually several days) fixation, the sieved samples are gently, but thoroughly, washed in fresh water. This removes the formalin and salt, preventing the former from dissolving the shells of delicate molluscs. The washing stage should be carried out in well-ventilated areas (e.g. in the open air) or fume cupboards. Formaldehyde respirators and goggles can also be worn if fumes present a problem. The samples are then preserved in 80% (v/v) alcohol.

The specimen-rich initial washing fractions are sorted into phyla under the dissection microscope. The remaining fractions are sorted by eye using a well-lit white tray. Pliable stork-bill forceps should be used throughout to prevent damage to the delicate forms, such as thin-shelled molluscs.

For each quantitative replicate, all specimens are enumerated and identified to the most advanced level possible relative to the available taxonomic literature and the time-scale for the completion of the project.

Sampling and Extraction Limitations and Resolutions

As there is no ideal sampling device, workers must endeavour to choose the best available for their purposes. In some cases, the sampler and/or sampling method chosen as 'standard' may have to be supplemented by a more specific one (e.g. to properly collect deep-burrowing invertebrates). The size and number of samples and sieve mesh size must also be considered.

The sampling protocol may have to be modified to take account of local conditions. For example, in that presented above, it is essential that a copious supply of running seawater is available for the sieving procedure. This is readily available on board research vessels, but will usually not be present at or near the site of an intertidal study.

There are two main solutions to the problem. First, the samples can be collected from the field situation and transported intact to a facility supplied with running seawater or to a location suitable for the ready extraction of seawater (e.g. by portable two-stroke water pump). Second, sieving may be carried out *in situ* by 'puddling' the sieve at the water's edge or in a purposely excavated water-filled hole. Sometimes it may be possible to use a portable water pump on site. The pump can be set up at the water's edge and the samples elutriated from a large plastic bin. To avoid contamination of the samples by other benthic animals (e.g. surface-dwelling crustaceans) the inlet hose should not rest on the seabed itself and it should be fitted with a suitable mesh filter. The pump will, however, need to be regularly moved in response to tidal movements. This can be avoided if the pump is placed in a small shallow-draught dinghy and sieving takes place on an incoming tide. If an inflatable dinghy is employed, the water pump and its hot exhaust should not be allowed to come in contact with the sides.

Extensive intertidal mudflats can be particularly difficult to sample. Where the mud is so soft that there is a danger of workers becoming stuck, a hovercraft may be necessary. Alternatively, the collecting method could be altered to allow sampling to take place at high tide using a small boat or inflatable dinghy. Invaluable guidance on safety while carrying out all types of marine biological sampling can be found in Baker and Wolff (1987).

Bibliography

Ankar, S. (1977) Digging profile and penetration of the Van Veen grab in different sediment types. *Contributions from the Åsko Laboratory* 16, 1-22.

Bachelet, G. (1990) The choice of a sieving mesh size in the quantitative assessment of marine macrobenthos: a necessary compromise between aims and constraints. *Marine Environment Research* 30, 21-35.

Baker, J.M. and Wolff, W.J. (1987) *Biological Surveys of Estuaries and Coasts.* Cambridge University Press, Cambridge.

Bamber, R.N. (1995) Sampling and data analysis: what did you want to know? *Porcupine Newsletter* 6, 63-68.

Barnett, P.R.O., Watson, J. and Connelly, D. (1984) A multiple corer for taking virtually undisturbed samples from shelf, bathyal and abyssal sediments. *Oceanologica Acta* 7, 399-408.

Beukema, J.J. (1974) The efficiency of the Van Veen grab compared with the Reineck Box sampler. *Journal du Conseil Permanent International pour l'Exploration de la Mer* 35, 319-327.

Birkett, L. (1958) A basis for comparing grabs. *Marine Biology, Berlin* 23, 202-207.

Bishop, J.D.D. and Hartley, J.P. (1986) A comparison of the fauna retained on 0.5 mm and 1.0 mm meshes from benthic samples taken in the Beatrice Oilfield, Moray Firth, Scotland. *Proceedings of the Royal Society of Edinburgh* 91B, 247-262.

Blake, J.A. and Hilbig, B. (1994) Dense infaunal assemblages on the continental slope off Cape Hatteras, North Carolina. *Deep-Sea Research II* 41, 875-899.

Boase, N.A. and Waller, R.R. (1994) The effect of propylene glycol on ethanol concentrations determined by density measurement. *Collection Forum* 10, 41-49.

Bros, W.E. and Cowell, B.C. (1987) A technique for optimising sample size (replication). *Journal of Experimental Marine Biology and Ecology* 114, 63-71.

Burd, B.J., Nemec, A. and Brinkhurst, R.O. (1990) The development and application of analytical methods in benthic marine infaunal studies. *Advances in Marine Biology* 26, 169-247.

Chang, W.Y.B. and Winnell, M.H. (1981). Comment on the fourth-root transformation. *Journal of the Fisheries Research Board of Canada* 38, 126-127.

Christie, N.D. (1976a) The efficiency and effectiveness of a diver-operated suction sampler on a homogeneous macrofauna. *Estuarine and Coastal Marine Science* 4, 687-693.

Christie, N.D. (1976b) A numerical analysis of the distribution of a shallow sublittoral sand macrofauna along a transect at Lamberts Bay, South Africa. *Transactions of the Royal Society of South Africa* 42, 149-171.

Cuff, W. (1980) Reply to: Comment on optimal survey design. *Journal of the Fisheries Research Board of Canada* 37, 297.

Cuff, W. and Coleman, N. (1979) Optimal survey design: lessons from a stratified random sample of macrobenthos. *Journal of the Fisheries Research Board of Canada* 36, 351-361.

Downing, J.A. (1979) Aggregation, transformation, and the design of benthos sampling programs. *Journal of the Fisheries Research Board of Canada* 36, 1454-1463.

Downing, J.A. (1980a) Correction to: Aggregation, transformation, and the design of benthos sampling programs. *Journal of the Fisheries Research Board of Canada* 37, 1333.

Downing, J.A. (1980b) Precision vs. generality: a reply. *Journal of the Fisheries Research Board of Canada* 37, 1329-1330.

Downing, J.A. (1981) How well does the fourth-root transformation work? *Journal of the Fisheries Research Board of Canada* 38, 127-129.

Eleftheriou, A. and Holme, N.A. (1984) Macrofauna techniques. In: Holme, N.A. and McIntyre, A.D. (eds) *Methods for the Study of Marine Benthos. IBP Handbook 16* (2nd edn), Blackwell Scientific Publications, Oxford, pp. 140-216.

Elliott, J.M. (1977) Some methods for the statistical analysis of samples of benthic invertebrates (second edition). *Scientific Publications of the Freshwater Biological Association* 25, 1-156.

Elliott, J.M. and Tullett, P.A. (1978) A bibliography of samplers for benthic invertebrates. *Occasional Papers, Freshwater Biological Association* 4, 1-61.

Fauchald, K. (1977) The polychaete worms. Definitions and keys to the orders, families and genera. *Science Series, Natural History Museum of Los Angeles County* 28, 1-188.

Gage, J.D. (1975) A comparison of the deep-sea Epibenthic Sledge and Anchor-box Dredge Samplers with the Van Veen grab and hand coring by diver. *Deep-Sea Research* 22, 693-702.

Gallardo, V. A. (1965) Observations on the biting profiles of three 0.1m^2 bottom samplers. *Ophelia* 2, 319-322.

Gamble, J.C. (1984) Diving. In: Holme, N.A. and McIntyre, A.D. (eds) *Methods for the Study of Marine Benthos. IBP Handbook 16* (2nd edn), Blackwell Scientific Publications, Oxford, pp. 99-139.

George, J.D. and Hartmann-Schröder, G. (1985) Polychaetes: British Amphinomida, Spintherida and Eunicida. *Synopses of the British Fauna, New Series* 32, 1-221.

Giere, O. and Pfannkuche, O. (1982) Biology and ecology of marine Oligochaeta, a review. *Oceanography and Marine Biology* 20, 173-308.

Grassle, J.F. and Maciolek, N.J. (1992) Deep-sea species richness: regional and local diversity estimates from quantitative bottom samples. *American Naturalist* 139, 313-341.

Green, R.H. (1978) Optimal impact study design and analysis. In: Dickson, K.L., Cairns, J., Jr. and Livingston, R.J. (eds) *Biological Data in Water Pollution Assessment: Quantitative and Statistical Analyses. American Society for Testing and Materials Scientific and Technical Publication* 652, pp. 3-28.

Green, R.H. (1979) *Sampling Design and Statistical Methods for Environmental Biologists.* John Wiley and Sons, New York.

Green, R.H. (1980) Comment on optimal survey design. *Journal of the Fisheries Research Board of Canada* 37, 296-297.

Harris, R.H. (1978) Biodeterioration. *Biology Curators' Group Newsletter* 8, 3-12.

Harris, R.H. (1992) Zoological preservation and conservation techniques. I. Fluid preservation. *Journal of Biological Curation* 1, 5-24.

Hartley, J.P. (1989) Methods for monitoring offshore macrobenthos. *Marine Pollution Bulletin* 13, 150-154.

Hartley, J.P. and Dicks, B. (1987) Macrofauna of subtidal sediments using remote sampling. In: Baker, J.M. and Wolff, W.J. (eds) *Biological Surveys of Estuaries and Coasts.* Cambridge University Press, Cambridge, pp. 106-130.

Hartley, J.P., Dicks, B. and Wolff, W.J. (1987) Processing sediment macrofauna samples. In: Baker, J.M. and Wolff, W.J. (eds) *Biological Surveys of Estuaries and Coasts.* Cambridge University Press, Cambridge, pp. 131-139.

Hessler, R.R. and Jumars, P.A. (1974) Abyssal community analysis from replicate box cores in the central North Pacific. *Deep-Sea Research* 21, 185-209.

Hines, A.H. and Comtois, K.L. (1985) Vertical distribution of infauna in sediments of a subestuary of central Chesapeake Bay. *Estuaries* 8, 296-304.

Hiscock, K. (1987) Subtidal rock and shallow sediments using diving. In: Baker, J.M. and Wolff, W.J. (eds) *Biological Surveys of Estuaries and Coasts.* Cambridge University Press, Cambridge, pp. 198-237.

Höisaeter, T. and Matthiesen, A.-S. (1979) *Report on some Statistical Aspects of Marine Biological Sampling Based upon a UNESCO-Sponsored Training*

Course in Sampling Design for Marine Biologists. University of San Carlos, Cebu City.

Holme, N.A. and McIntyre, A.D. (1984) *Methods for the Study of Marine Benthos. IBP Handbook 16,* 2nd edn. Blackwell Scientific Publications, Oxford.

Johnson, R.G. (1967) The vertical distribution of the infauna of a sand flat. *Ecology* 48, 571-578.

Jumars, P.A. (1978) Spatial autocorrelation with RUM (Remote Underwater Manipulator): vertical and horizontal structure of a bathyal benthic community. *Deep-Sea Research* 25, 589-604.

Kingston, P. F. and Riddle, M. J. (1989) Cost effectiveness of benthic faunal monitoring. *Marine Pollution Bulletin* 20, 490-496.

Kutty, M.K. and Desai, B.N. (1968) A comparison of the efficiency of the bottom samplers used in benthic studies off Cochin. *Marine Biology, Berlin* 1, 168-171.

Levi, H.W. (1966) The care of alcoholic collections of small invertebrates. *Systematic Zoology* 15, 183-188.

Lie, U. (1968) A quantitative study of benthic infauna in Puget Sound, Washington, USA in 1963-1964. *Fiskeridirektoratets Skrifter (HavundersØkelser)* 14, 223-556.

Lie, U. and Pamatmat, M.M. (1965) Digging characteristics and sampling efficiency of the 0.1 m^2 Van Veen grab. *Limnology and Oceanography* 10, 379-385.

Lincoln, R.J. and Sheals, J.G. (1979) *Invertebrate Animals Collection and Preservation.* British Museum (Natural History), London/Cambridge University Press, Cambridge.

McIntyre, A.D. (1969) Ecology of marine meiobenthos. *Biological Reviews (and Biological Proceedings) of the Cambridge Philosophical Society* 44, 245-290.

McIntyre, A.D., Elliott, J.M. and Ellis, D.V. (1984) Introduction: Design of sampling programmes. In: Holme, N.A. and McIntyre, A.D. (eds) *Methods for the Study of Marine Benthos. IBP Handbook 16,* 2nd edn. Blackwell Scientific Publications, Oxford, pp. 1-26.

McIntyre, A.D. and Warwick, R.M. (1984) Meiofauna techniques. In: Holme, N.A. and McIntyre, A.D. (eds) *Methods for the Study of Marine Benthos. IBP Handbook 16,* 2nd edn. Blackwell Scientific Publications, Oxford, pp. 217-244.

Mackie, A.S.Y. (1994) Collecting and preserving polychaetes. *Polychaete Research* 16, 7-9.

Mackie, A.S.Y. and Chambers, S.J. (1990) Revision of the type species of *Sigalion*, *Thalenessa* and *Eusigalion* (Polychaeta: Sigalionidae). *Zoologica Scripta* 19, 39-56.

Mackie, A.S.Y., Oliver, P.G. and Rees, E.I.S. (1995) Benthic biodiversity in the southern Irish Sea. *Studies in Marine Biodiversity and Systematics from the National Museum of Wales. BIOMÔR Reports* 1, 263 pp.

Ockelmann, K.W. (1989) Suggested procedure for fixation and preservation of benthos samples (Metazoans). In: Heip, C. and Niermann, U. (eds) *Taxonomy of North Sea Benthos. Proceedings of a Workshop organised in Helgoland 8-12 February 1988.* Delta Institute for Hydrobiological Research, Yerseke, The Netherlands, p. 4.

Paterson, G.L.J., Thurston, M.H., Gage, J.D., Lamont, P.M. and Bett, B. J. (1994) Patterns of polychaete assemblabe structure from the abyss: some preliminary observations from NE Atlantic abyssal plains. *Polychaete Research* 16, 16-19.

Pleijel, F. (1990) A revision of the genus *Sige* Malmgren (Polychaeta: Phyllodocidae). *Zoological Journal of the Linnean Society of London* 98, 161-164.

Pleijel, F. and Dales, R.P. (1991) Polychaetes: British Phyllodocoideans, Typhloscolecoideans and Tomopteroideans. *Synopses of the British Fauna, New Series* 45, 1-202.

Rees, E.I.S., Allen, P.L. and Coppock, J. (1994) Representative replication for sediment benthos monitoring: Application of varied strategies in the Irish Sea. *Porcupine Newsletter* 5, 225-233.

Rees, H.L. (1984) A note on mesh selection and sampling efficiency in benthic studies. *Marine Pollution Bulletin* 15, 225-229.

Rees, H.L., Moore, D.C., Pearson, T.H., Elliott, M., Service, M., Pomfret, J. and Johnson, D. (1990) Procedures for the monitoring of marine benthic communities at UK sewage sludge disposal sites. *Scottish Fisheries Information Pamphlet* 18, 1-78.

Rees, H.L., Heip, C., Vincx, M. and Parker, M.M. (1991) Benthic communities: use in monitoring point-source discharges. *ICES, Copenhagen, Techniques in Marine Environmental Sciences* 16.

Reish, D.J. (1959) A discussion of the importance of screen size in washing quantitative marine bottom samples. *Ecology* 40, 307-309.

Reish, D.J. (1980) Use of polychaetous annelids as test organisms for bioassay experiments. In: Buikema, A.L. and Cairns, J. (eds) *Aquatic Invertebrate Bioassays. American Society for Testing and Materials Scientific and Technical Paper* 715, pp. 140-154.

Riddle, M.J. (1984) *Offshore Benthic Monitoring Strategies.* PhD Thesis, Heriot-Watt University, Edinburgh.

Riddle, M.J. (1989a) Precision of the mean and the design of benthos sampling programmes: caution advised. *Marine Biology, Berlin* 103, 225-230.

Riddle, M.J. (1989b) Bite profiles of some benthic grab samplers. *Estuarine, Coastal and Shelf Science* 29, 285-292.

Rosenberg, D.M. (1978) Practical sampling of freshwater macrozoobenthos: a bibliography of useful texts, reviews and recent papers. *Technical Reports, Fisheries and Marine Service, Canada* 790, 1-15.

Rumohr, H. (1990) Soft bottom macrofauna: collection and treatment of samples. *ICES, Copenhagen, Techniques in Marine Environmental Sciences* 8.

Sanders, H.L. (1960) Benthic studies in Buzzards Bay. III. The structure of the soft-bottom community. *Limnology and Oceanography* 5, 138-153.

Scherba, S. and Gallucci, V.F. (1976) The application of systematic sampling to a study of infauna variation in a soft substrate environment. *Fishery Bulletin, National Oceanic and Atmospheric Administration* 74, 937-948.

Schwinghamer, P. (1981) Characteristic size distributions of integral benthic communities. *Canadian Journal of Fisheries and Aquatic Science* 38, 1255-1263.

Shirayama, Y. and Horikoshi, M. (1982) Vertical distribution of smaller macrobenthos and larger meiobenthos in the sediment profile in the deep-sea system of Suruga Bay (central Japan). *Journal of the Oceanographical Society of Japan* 38, 273-280.

Smaldon, G. (1978) In praise of soda-water. *Biology Curators' Group Newsletter* 2, 18-19.

Smaldon, G. and Lee, E.W. (1979) A synopsis of methods for the narcotisation of marine invertebrates. *Information Series. Royal Scottish Museum (Natural History)* 6, 1-96.

Smith, W. (1978) Environmental survey design: a time series approach. *Estuarine and Coastal Marine Science* 6, 217-224.

Taylor, L.R. (1980) New light on the variance/mean view of aggregation and transformation: comment. *Journal of the Fisheries Research Board of Canada* 37, 1330-1332.

Taylor, W.D. (1980) Comment on: Aggregation, transformation, and the design of benthos sampling programs. *Journal of the Fisheries Research Board of Canada* 37, 1328-1329.

Tenore, K.R., Browne, M.G. and Chesney, E. (1974) Polyspecies aquaculture systems: the detrital trophic level. *Journal of Marine Research* 32, 425-432.

Tyler, P. and Shackley, S.E. (1978) Comparative efficiencies of Van Veen and Smith-McIntyre grab samples. *Estuarine and Coastal Marine Science* 6, 439-445.

Vézina, A.F. (1988) Sampling variance and the design of quantitative surveys of the marine benthos. *Marine Biology, Berlin* 97, 151-155.

Warwick, R.M. (1984) Species size distributions in marine benthic communities. *Oecologia* 61, 32-41.

Warwick, R.M. (1993) Environmental impact studies on marine communities: pragmatical considerations. *Australian Journal of Ecology* 18, 63-80.

Warwick, R.M. and Clarke, K.R. (1993) Comparing the severity of disturbance: a meta-analysis of marine macrobenthic community data. *Marine Ecology Progress Series* 92, 221-231.

Westerberg, J. (1978) Benthic community structure in the Åland archipelago (N Baltic) represented by samples of different sizes). *Kieler Meeresforschungen, Sonderheft* 4, 53-60.

Westheide, W. (1990) Polychaetes: Interstitial families. *Synopses of the British Fauna (New Series)* 44, 1-152.

Wigley, R.L. (1967) Comparative efficiencies of Van Veen and Smith-McIntyre grab samples as revealed by motion pictures. *Ecology* 48, 168-169.

Wildish, D.J. (1978) Sublittoral macro-infaunal grab sampling reproducibility and cost. *Technical Reports, Fisheries and Marine Service, Canada* 770, 1-14.

Wolff, W.J. (1987) Flora and macrofauna of intertidal sediments. In: Baker, J.M. and Wolff, W.J. (eds) *Biological Surveys of Estuaries and Coasts.* Cambridge University Press, Cambridge, pp. 81-105.

Word, J.Q. (1976) Biological comparison of grab sampling devices. *Southern California Coastal Water Research Project. Annual Report (1976)*, pp. 189-194.

Burrowing Mega- and Macrofauna from Marine Sediments: the Minor Groups

22

J. DOUGLAS MCKENZIE

Scottish Association for Marine Science, P.O. Box 3, Oban, Argyll PA34 4AD, UK.

The Role of Burrowing Mega- and Macrofauna in Marine Sediments

Size is the critical feature defining the importance of the 'minor' groups to sediment processes. Many of the organisms that form the subject of this section are relatively huge compared with the majority of infaunal marine organisms. Some species frequently reach 30 cm in length and can be in excess of 1 m. Such large organisms produce considerable bioturbation in marine sediments. Burrow depth can exceed 1 m and lateral tunnelling can extend for several metres.

The precise influence of the burrower will depend on its feeding habit. Surface deposit-feeders alter the surrounding top layers of sediment while suspension feeders produce local particle deposition. Sub-surface deposit feeding and the actual burrowing process produces considerable sediment reworking. This will affect the local nutrient availability as 'locked' nutrients are released from the sediments. It may also release pollutants such as toxins and radionucleotides that had previously been buried in the sediment. Burrows will increase oxygenation of the sediment to depths well below the normal redox potential discontinuity layer, with species that actively pump water through their burrows having the greatest effect.

These large bioturbators will thus considerably alter the chemistry and physical structure of the sediment, producing a heterogeneity that is reflected in altered faunal community structure. The activities of holothurians, echiurans and enteropneusts have been shown to alter community structure. Sediment disturbance by burrowers can be detrimental to some species, while others exploit the opportunities it provides. Alterations to community structure act at all scales: from other macrofaunal elements, through meiofauna to bacterial populations. Bacterial populations associated with infaunal invertebrates and their burrows are usually more diverse than those found in nearby, undisturbed sediments.

While there is a general recognition of the potential importance of megafaunal burrowers on sediment processes, hard facts are rare. This is due to a combination of factors, but under-sampling of burrowing megafauna is a major contributor to our ignorance. Increasing use of remote and diving techniques has begun to reveal the severe limitations of conventional techniques in sampling large burrowers. Density obviously multiplies the importance of individual size and biomass measurements exceeding several kg m^{-2} have been obtained for such megafauna. Under sampling of such large organisms can completely distort ecosystem understanding and is likely to be important in community studies of the effects of pollution.

Not all the members of this diverse collection of animals can be classed as megafauna (large enough to be seen in an underwater photograph). Many species are essentially macrofaunal, i.e. they are retained by a 1 mm sieve. A few are even smaller than this, such as the synaptid holothurian *Leptosynapta minuta,* and recently settled post-larvae are also very small. Generally, intertidal species are usually smaller than subtidal species. The density of these smaller burrowers, rather than their size, has a direct effect on sediments; denser populations having a greater impact. Qualitatively, smaller burrowers have the same effect as the larger burrowers, but, even at high densities, they may not achieve the quantitative impact of larger burrowers, since they do not burrow as deeply.

The similarities between the different groups considered here are due, in large part, to convergent evolutionary pressures produced by a burrowing lifestyle, rather than phylogeny. Almost all are exclusively marine, though some burrowing anemones can be found in brackish water. All of the species in these groups have cylindrical bodies, usually rounded at the base, and many have an extendable feeding structure. Six groups are covered here.

Burrowing Sea Anemones

Many anthozoans partially cover themselves with silt and gravel but the true burrowing forms are in the minority. Two separate groups exist: members of the order *Ceriantharia*, which live in lined tubes, and true anemones of the tribe *Athenaria*, which is distinguished from other anemones by the lack of a muscular, basal disc. Both groups suspension-feed or attack larger prey which encounter their tentacles. Some cerianthid species are very large and may be significant in sediment processes. PHYLUM *CNIDARIA*

Echiurans (Spoon Worms)

There are approx. 130 species worldwide. Most live in burrows or under rocks in soft sediments. Most surface deposit-feed by use of a greatly extendable proboscis, though some are mucus-net suspension-feeders. Larger species can exceed 50 cm in body length (excluding the proboscis, which may be some metres in length). Such species form deep burrows and can be difficult to catch. They are important but understudied bioturbators. Large echiurans have frequently been photographed in the deep sea. They are usually common in coastal waters and some species live in the intertidal region. PHYLUM *ECHURIA*

Priapulids

One of the smallest known phyla, with probably less than ten species, which are in a state of taxonomic confusion. Some species are nevertheless widespread and sometimes locally common. Small- to medium-sized (< 15 cm), they can be significant bioturbators when populations reach high densities. Usually described as predators, there are also reports of their filter-feeding. They are usually found on the continental shelf of temperate and polar seas, though a minute tropical species has recently been described. PHYLUM *PRIAPULIDA*

Sipunculids (Peanut Worms)

This group has over 300 described species with a worldwide distribution from a wide range of habitats. Sipunculids are usually found in soft sediments, although many species are important borers, particularly of coral reefs. Body size in adults ranges from tiny (3 mm) to very large (> 50 cm), depending on species. They are believed to be mostly sub-surface deposit-feeders. PHYLUM *SIPUNCULIDA*

Burrowing Holuthurians (Sea Cucumbers)

There are just under 1500 species of holothurians known worldwide, over half of which are burrowers. Of these, many are suspension-feeders (particularly the dendrochirotes), although others are surface or subsurface deposit-feeders (notably the two apodan orders). Burrowing species range from small, delicate species to others in excess of 50 cm. What little is known about the bioturbating activities of holothurians suggests that they have an important impact on the ecology of the surrounding sediments. They may be found in all marine habitats. PHYLUM *ECHINODERMATA*

Enteropneusts (Acorn Worms)

There are fewer than 100 described species of enteropneusts, most of which are burrowers. Although they are known mainly from the intertidal region, there is increasing recognition of their importance in deep-sea sediments as well. They are usually moderately large species, ranging from 5 to 20 cm, but some species are enormous with recorded lengths of over 2 m. They are either sub-surface deposit-feeders or suspension-feeders. PHYLUM *HEMICHORDATA*

Sampling Techniques

There are no established, specialized sampling protocols for use with any of these groups of animals. The broad range of sizes and habitats necessitates an equally broad range of techniques to obtain them. In most cases, biodiversity surveys are unlikely to be specifically targeted at these groups and they are likely to be encountered during more general benthic surveys, aimed at sampling the entire fauna. With the exception of some of the megafauna, the general methods used for sampling benthos are applicable to these groups, but more attention has to be paid to sample sorting and processing. A good guide to general benthic sampling techniques is Holme and McIntyre (1984) and it is not intended to exhaustively review them here. The problems in sampling burrowing megafauna tend to be: obtaining specimens; producing an accurate quantification of density and biomass; and obtaining specimens in good condition for ease of subsequent identification.

Specimen Capture

Intertidal organisms

All of the groups dealt with here may be encountered in the intertidal region and studies of intertidal forms are often the most detailed sources of information concerning their biology. Enteropneusts can be located by looking for evidence of their characteristic faecal mounds (Ruppert and Fox, 1988). However, most groups are rarely discernible before sampling. Collecting intertidal organisms is relatively straightforward and rarely requires more than a sieve, a bucket, a shovel and some labour. For quantitative studies, thin metal frames can be pushed into the sediment to the required depth and all the sediment within the frame removed prior to sieving. Because many of the species are deep burrowers, even in the intertidal region, sediment should ideally be removed to a depth of 50 cm. Intertidal areas are very vulnerable to disturbance, so it is important to keep sampling to the necessary minimum.

Sieving the sediment is essential to provide quantitative information and is also the best way of isolating species of these groups from the sediment. However, it is worthwhile to initially place spadefuls of the sediment to one side and allow it to break apart gently either by its own weight or by tapping it with the spade. Any burrows will present lines of weakness in the sediment which will tend to break apart around them. Any megafauna exposed in this way may then be removed gently by carefully washing away the sediment from around the animal. **Removal of the animal by pulling it from the sediment before it is exposed should never be attempted**, as delicate species such as synaptid holothurians and enteropneusts will almost certainly break. Procuring intact animals will greatly assist subsequent identification. Sieving should also be done as gently as possible to help prevent breakage of specimens. Sieves with a mesh size of 1 mm will be too fine for use on coarser sediments and it may be best to use two seives in series: an initial coarse one of 5 or 10 mm and a finer one underneath. Obtaining sufficient water for sieving can be a problem, but it should be brought to the sieve (in buckets) and not vice versa. Care should be taken when using streams that run down the beach as sources of water, as even brackish water will be of sufficiently low salinity to cause extreme distortion and death of many of the species. Sieving should not be done in rock pools, as this may lead to damage of the rock pool community.

Because of the limited time available for sampling in the intertidal region, it is best to concentrate on obtaining specimens and then transporting them live to the laboratory or suitable sorting point. Samples should be placed in plastic bags with plenty of cool sea water and transported in insulated 'cool bins'. If there will be a significant delay between sample collection and sorting, then filling the bags with oxygen before closing will aid survival of animals. Specimens should be kept in the dark and as cool as possible.

Subtidal organisms

Sampling sublittoral and deeper habitats requires the use of boats, submersibles or divers. Sampling programmes are inevitably compromises, and aim to extract the largest amount of useful data within the available resources rather than to do exhaustive and definitive studies. For biodiversity studies, it is best to use as many different types of sampling methods as often as possible, rather than rely on a single technique. For population studies, the opposite applies, as it is best to have as much directly comparable data as possible. Adding to these complications are the mundane practicalities of sampling at sea. Much of the time the 'best' techniques may be useless because of weather, sediment type or the non-availability of appropriate gear. This makes it difficult to produce a rigid sampling protocol, as obtaining samples must take precedence over the subsequent accuracy of quantification. Sampling the deep-burrowing megabenthos can sometimes be so difficult that desperate measures are required to obtain any specimens, let alone quantifiable data. There is no single, optimal technique suitable for the groups of animals covered in this section. As the size of the organism decreases, so the efficiency standard quantitative methods increases. However, quantitative methods usually give an inaccurate assessment of the importance of less common species, regardless of their size. Table 22.1 shows the range of techniques available as well as their respective advantages and disadvantages.

Table 22.1. Comparison of sampling methods for obtaining burrowing megafauna for subtidal sediments.

Technique	Type	Advantages	Disadvantages
Diving	Both	Relatively cheap, all bottom-type capability, very targeted	Severe time and depth limitations, limited penetration of sediment
Submersibles	Both	Good time and depth capabilities, large recording capacity, all bottom-type capability, very targeted	Extremely expensive, limited penetration of sediment
Remote surveillance	Both	Excellent time and depth capabilities, very targeted	May be expensive, no specimen retrieval capability
Trawls and dredges	NQ	Cover large area, relatively cheap, good depth and operational capability, can be cheap	Limited sediment penetration, possible damage to specimens
Anchor dredge	Semi-Q	Good sediment penetration, cheap	Depth of use limited, small area coverage
Grabs	Q	Quantitative, reasonable sediment penetration, relatively cheap	Sediment type very limited, small area coverage
Corers	Q	Good sediment penetration, burrow structure intact, good specimen preservation, good depth range	Requires large vessel, expensive, limited area coverage

Type column refers to the ability of the technique to quantify population density. NQ non-quantitative, Q quantitative.

Diving is an excellent tool for shallow-water, macrofaunal burrowers which surface-feed. For example, it has been used to good effect (in combination with other techniques) to study a large echiuran species (Hughes *et al.*, 1996). Diving can be used to investigate any type of sea bottom and is thus particularly useful on harder ground, where conventional sampling gear would be difficult to employ. Flexibility is a key feature of diving capability. Divers can undertake survey work, take photographs and videos, place equipment around burrows and directly capture specimens. Although initial training and equipment are expensive, diving is a relatively cheap sampling method compared with use of a boat. Using compressed air, useful diving is restricted to water depths of between 30-40 m. Although divers can go deeper, their no-stop bottom time becomes very short and the value of their observations decreases with the onset of nitrogen narcosis. Mixed-gas techniques can increase the depth range and bottom time but also increase the expense and complexity of diving operations. Divers are good at surveying epifauna but burrowers are often overlooked. This can occur even with megafauna that surface-feed. Dendrochiriote holothurians, for instance, only feed for part of the year and are impossible for divers to locate at other times.

Extraction of burrowing megafauna by divers is not easy. Many species will retreat into their burrows or into the sediment at the mere approach of a diver. Large suspension-feeders such as dendrochirote holothurians and cerianthid anemones may be extracted by firmly grasping the feeding structure and pulling. Additional leverage can be had by plunging a knife into the sediment below the specimen and levering upwards. However, these techniques have a low success rate, and usually result in the removal of only the autotomized feeding structure or a knife through the animal's body.

For synaptid holothurians in soft mud an unusual sampling technique is to have the diver roll around in the mud. The synaptids will adhere to the material surface of the wet or dry suit (rubber or plastic-surfaced suits do not work well) and can be carefully removed once the diver has left the water.

Punch cores (Gage, 1975) are mostly useful for collecting meiofauna but can sometimes catch shallow-burrowing macrofauna, particularly when these are common.

Suction devices, either hand-held 'slurp' guns or gas-operated suction samplers, can be used to extract specimens from near their burrow mouths or to excavate them (Holme and McIntyre, 1984). These techniques usually have a low success rate (Hughes, pers. comm.). Powerful suction devices may also damage more delicate specimens.

Manned submersibles can usually dive to greater depths and have superior endurance, otherwise their operational capabilities and disadvantages are very similar to divers. However, they cost considerably more, both initially and operationally.

There are a variety of remote sensing techniques useful for assessing the likely presence, distribution and bioturbation potential of burrowing megafauna. These include various camera and video systems, remote-operated vehicles (ROVs) and echo-sounding techniques, although none of these can retrieve specimens intact. As with divers and submersibles, they provide no information on burrowing macrofauna if they are not surface-feeding or showing evidence of their presence. They can, however, record benthic events over very long periods. *In situ* cameras are particularly useful, as individuals near the camera may soon habituate to its presence and behave normally. Time-lapse 'Bathysnap' cameras (Lampitt and Burnham, 1983) have shown that megafaunal burrowers, such as echiurans are more common in the deep sea than had previously been thought (Lampitt, 1985). Echo-sounding techniques such as 'RoxAnn' (Chivers *et al.*, 1990) can detect bioturbation traces which are characteristic of certain types of burrowing organisms and can be used to identify quickly areas for subsequent sampling likely to contain megafaunal burrowers. Some animals may be identified from photographs and video information. The colour and number of dendrochirote tentacles, the shape of echiuran proboscises and the characteristic feeding and defecation traces left on the sediment surface can sometimes give specific identification.

Dredges (Holmes and McIntyre, 1984) are open-mouthed, heavy metal frames with a net attached to the rear designed to at least partially dig into the sediment. They are the main method for sampling megafauna and, though best in collecting epifauna, they are also useful for obtaining infaunal organisms. They are either non- or semi-quantitative but have considerable advantages over grabs. They can be used in much rougher weather, sample a greater area more quickly and are more likely to catch fast-moving burrowers than grabs. They are also cheaper and can be used on a wider range of bottom types. Most dredges only sample the first few centimetres of sediment and so are restricted to capturing the smaller burrowers living near the surface. Deep-burrowing megafauna are usually only represented by their severed anteriors. The anchor dredge, however, is designed to dig deeply into the sediment and recover a large 'bite' of sediment - usually at least 20 cm in depth. It is considered semi-quantitative and is an excellent piece of equipment for capturing deep-burrowers. Its operation is limited to relatively shallow water (< 100m) and on soft muds, as its sediment penetration decreases with increasing sediment stiffness. Deep-sea anchor dredges do exist (Sanders *et al.*, 1965), but these have more limited sediment penetration. With dredges, care should be taken to examine the dredge frame and net as well as the contents. Synaptid holothurians are frequently found attached to the netting, particularly if the contents have been 'washed' during retrieval.

There are a large number of grab designs (Holme and McIntyre, 1984) but all work on the basis of metal jaws coming together to bite the sediment and capture it within the closed jaws. They are the method of choice for collecting quantitative samples but the limited depth of sediment they can penetrate (usually less than 15 cm) and the small area they can sample (0.1-0.2 m^2) means that they are usually poor at catching deep-burrowing species. Larger grabs do exist (Hartmann, 1955) but they are much heavier, making them potentially dangerous to operate. Grabs are restricted to use in soft sediments and their penetration and closing efficiency markedly drops off with increasing sediment stiffness. Ground with a mixture of soft sediments and cobbles or rocks is difficult to work as the stones often catch in the jaws resulting in the sediment being lost from the grab as it is hauled back on board the vessel.

Coring devices are open ended tubes or boxes that are pushed into the sediment either by external force or through the action of their own weight. The material is held in the corer during retrieval either by the cohesive properties of the cored sediment or by a core retainer which shuts off the core after it has penetrated the sediment. Corer sizes vary from simple tubes which can be operated by divers up to the large box corers used in deep-sea research (Gage and Tyler, 1991). The larger box corers are excellent for sampling burrowing megafauna. These can penetrate up to 50 cm into soft sediments and return a relatively undisturbed sample to the surface. Burrows are often recovered intact and, while some of the deepest burrowers can still evade capture (Jensen, 1992), sampling efficiency is high. The disadvantages with box corers are their size and weight, which necessitate the use of a large vessel to deploy them.

Some of the deep-burrowing megafauna are so difficult to sample using conventional gear that any potential source of specimens should be seized upon. The value of 'natural' sampling methods has been long understood. Forbes (1841) comments: "...Like the Haddock, the Cod also is a great naturalist; and he too, carries his devotion to our dear science so far as occasionally to die for its sake with a new species in its stomach, probably with a view to it being described and figured by some competent authority...". The stomachs of large, benthic-feeding fish do often contain megafauna, including burrowers - though they are rarely recovered in pristine condition. Beachcombing after a storm can be a useful source of information on local burrowing megafauna. Local people may have specialized techniques for extracting deep burrowers and dredging associated with civil engineering projects has also been recorded as a source of otherwise difficult-to-obtain specimens (Cutler, 1994).

Sediment Handling and Processing

Most of the techniques described result in a sample of sediment being brought up onto a boat, from which extraction of animals is necessary. The smaller representatives of the groups of animals considered here may turn up in samples processed using methods for meiofauna and small macrofauna (see Chapters 15-19, 21). Extraction of the larger macrofauna and the megafauna usually consists simply of washing the sediment away, while retaining the animals. For quantitative studies, the aim is to process as much sediment as possible in the available time. In studying biodiversity, the principal objective is usually to obtain specimens that can subsequently be accurately identified. As some of the species covered in this section are very delicate, the means of separating them from the sediment must be gentle, resulting in longer processing times. A further conflict is that longer processing times may result in damage to the specimens due to adverse local environmental conditions, such as temperature, oxygen availability, etc. Sediment processing will therefore be a compromise between time and specimen preservation pressures. The latter depends on the characteristics of individual species, making it impossible to produce a 'best practice' protocol for the species covered here.

To sample megafauna which have low population densities, large volumes of sediment need to be processed. This has to be done quickly, but gently. The usual methods of processing utilize large sieving hoppers or trays. The sediment is placed in these containers and the ship's seawater hoses are used to wash the sediment through the sieve. Samples usually come from sediments that are cooler than surface waters and this can lead to the animals dying from thermal stress. This is a particular problem with deep-sea specimens taken from tropical waters. Another problem is the presence of low-salinity water at the surface of the sea, which frequently happens near river outflows or after heavy rain. Using low-salinity water to process the sediments will quickly result in the death and distortion of delicate organisms such as apodous holothurians. In most cases it is not possible to circumvent these problems, as a ship's seawater system is usually restricted to drawing its supply from near the sea surface.

Once freed from the sediment, specimens should be rapidly placed into buckets with seawater of the appropriate temperature and salinity to minimize the effect of these problems. Anemones should not be kept together or with other organisms as they will attack and eat them. It is best not to process the sediment completely before removing the animals as this will lead to physical damage. Water should instead be played over the sediment for a few minutes and then allowed to drain away. Any animals exposed are then carefully removed with further gentle washing, if needed. This process should be repeated until all the sediment has been washed away. If the available seawater is of compatible temperature and salinity, then gentler (and longer) fluidization techniques may be used (e.g. in Sanders *et al.*, 1965).

Specimen Anaesthesia, Fixation and Preservation

A good, simple guide to techniques is Lincoln and Sheals (1979). Many of the groups covered here are considered difficult to identify. However, this is frequently because specimens have been poorly preserved or are incomplete. Colour photographs of the living specimens and notes on the sediment type they were found in, the associated fauna, depth, etc. may aid identification. All of these animals are highly contractile and many have a tendency to readily eviscerate or autotomize structures. Anterior feeding structures, such as tentacle number in holothurians, the proboscis shape in echiurans and the introvert of sipunculids, are often crucial to identifying them. These are also the structures most likely to be contracted into the body or lost during preservation. If time allows, anaesthetization prior to fixation is worthwhile; the aim being to preserve the organism in as relaxed a condition as possible, so that its feeding appendages are extended and the body wall not unnaturally thickened. Many of the organisms rapidly autolyse, so fixation should be as prompt a process as possible.

There are a number of anaesthetics in general use. A 7.5% (w/v) solution of magnesium chloride, dissolved in fresh water is popular, but slow to affect large animals. Ethyl alcohol (10% (v/v) final volume, made from absolute ethanol, not industrial methylated spirits (IMS)) added slowly to seawater in which the specimen is placed will produce satisfactory anaesthetization, although it must be added slowly to prevent contraction of the animals and discharge of nematocysts in the burrowing anemones (Manual, 1981). Propylene phenoxetol (1% v/v aqueous solution) will also give good results for all these groups of animals. Fixation should not be attempted until the specimens stop responding to stimuli. Before placing into the fixative, feeding structures should be grasped with a pair of tweezers to prevent them retracting when the animal is placed in the fixative.

There is a wide range of potential fixatives suitable for megafauna (Lincoln and Sheals, 1979). Many of these, however, should only be considered if the specimens are subsequently going to be used for histological studies. In practice, > 70% (v/v) ethanol and 5% (v/v) formalin in seawater are probably the two most practical and widely used fixatives. Formalin provides better cytological fixation and colour preservation than alcohol, but it has several disadvantages: it must be buffered to prevent dissolution of calcite structures which can be important in identification; it is much more unpleasant to use than ethanol; and, its use is increasingly being restricted on health and safety grounds. It is the method of choice for burrowing anemones, as it avoids discharge of the nematocysts (Manual, 1981), but it should be avoided for holothurians because it frequently leads to loss of the ossicles, which are important in identification. For this reason formalin fixation should be avoided if the specimens have not been identified to phylum. Alcohol is a convenient fixative, as it is also the most common preservation medium. However, it is more bulky than formalin, may be expensive, may require special import licences for certain countries and is flammable. In most cases, 24 h in the fixative should be sufficient, but large specimens may also be injected with fixative, if preservation of the internal organs is important, e.g. sipunculids (Stephen and Edmonds, 1972).

Both formalin (5-10% v/v aqueous solution) and alcohol (70-80% v/v IMS may be used) are good preservatives. However, specimens soaked in formalin are unpleasant to work with. Formalin is an unsuitable storage medium for holothurians, as even buffered formalin will eventually dissolve the ossicles, rendering the specimen unidentifiable. Alcohol may cause severe shrinkage of specimens, so, if they have been fixed in something other than alcohol, it may be worth increasing the concentration of the alcohol gradually. Alcohol will also extract most pigments from specimens, emphasizing the need for photographs or, at least, descriptions of the living specimens. An alternative preservative is propylene phenoxetol (1-2% v/v aqueous solution). This preserves colour well, is much more pleasant to use and is non-flammable. It must be stressed that propylene phenoxetol is only a preservative and prior fixation is essential.

Limitations of Existing Techniques

The major problem with burrowing megafauna is to catch them, and this problem appears intractable. Deep burrows require deep sediment penetration and this in turn requires heavy digging gear. As the gear weight increases so does the size of the ship needed to deploy it. Coring suction devices targeted onto burrows by cameras may be the best answer to actually catching specimens. The low population densities of some of these species (in numbers, if not biomass) also mean that they are under-represented in surveys. After capture, the techniques for processing these groups of animals are fairly straightforward, although they may be slow, if good preservation is required. However, in terms of understanding their biodiversity, probably the most important obstacle is the worldwide shortage of taxonomic expertise in these groups. For instance, there are fewer than 10 experts in holothurian taxonomy in the world. This slows down the description of new species and the writing of monographs and identification keys upon which the wider biological community relies.

Bibliography

Chivers, R.C., Emerson, N. and Burns, D.R. (1990) New acoustic processing for underway surveying. *Hydrographic Journal* 56, 9-17. [Review of acoustic profiling techniques.]

Cutler, E.B. (1994) *The Sipuncula: Their Systematics, Biology and Evolution*. Comstock, Ithaca. [Very good guide to the sipunculids with useful information on sampling.]

Forbes, E. (1841) *A History of British Starfishes and Other Animals of the Class Echinodermata*. Van Voorst, London. [Book of historical interest by one of the founders of marine ecology.]

Gage, J.D. (1975) A comparison of the deep sea epibenthic sledge and anchor-box dredge samplers with the van Veem grab and hand coring by divers. *Deep-Sea Research* 22, 693-702. [Paper on sampling methodology.]

Gage, J.D. and Tyler, P.A. (1991) *Deep-Sea Biology*. Cambridge University Press, Cambridge. [Good general guide to deep-sea biology with useful chapter on sampling methods.]

Hartmann, O. (1955) Quantitative survey of the benthos of San Pedro Basin, Southern California. Part 1. Preliminary results. *Allan Hancock Pacific Expedition* 19, 1-185. [Paper on deep-sea sampling techniques.]

Holmes, N.A. and McIntyre, A.D. (1984) *Methods for Studying the Marine Benthos*. Blackwell, Oxford. [Invaluable guide to all aspects of sampling benthos.]

Hughes, D.J., Ansell, A.D. and Atkinson, R.J.A. (1996) Sediment bioturbation by the echiuran worm *Maxmuellaria lankesteri* (Herdman) and its consequences for radionuclide dispersal in Irish Sea sediments. *Journal of Experimental Marine Biology and Ecology* (in press). [Paper on biology of megafaunal echiuran.]

Jensen, P. (1992) *Cerianthus vogti* Danielssen, 1890 (Anthozoa: Ceriantharia).A species inhabiting an extended tube system deeply buried in deep-sea sediments off Norway. *Sarsia* 77, 75-80. [Paper on sampling megafaunal sea anemone.]

Lampitt, R.S. (1985) Fast living on the ocean floor. *New Scientist* 105 (1445), 37-40. [News article on importance of burrowing megafauna in the deep sea.]

Lampitt, R.S. and Burnham, M.P. (1983) A free-fall, time lapse, camera and current meter systcm 'Bathysnap' with notes on the foraging behaviour of a bathyal decapod shrimp. *Deep-Sea Research* 30, 1009-1017. [Paper on camera system for photographing deep-sea organisms.]

Lincoln, R.J. and Sheals, J.G. (1979) *Invertebrate Animals: Collection and Preservation*. Cambridge University Press, London. [Very useful guide.]

Manual, R.L. (1981) *British Anthozoa*. Academic Press, London. [Identification guide with useful notes on collecting and preserving sea anemones.]

Ruppert, E.E. and Fox, R.S. (1988) *Seashore Animals of the Southeast*. University of South Carolina Press, Columbia. [General guide to the fauna of the east coast of the USA.]

Sanders H.L., Hessler, R.R. and Hampson, G.R. (1965) An introduction to the study of the deep-sea benthic faunal assembleges along the Gay Head-Bermuda transect. *Deep-Sea Research* 12, 845-867. [Paper giving details of deep-sea sampling methods.]

Stephen A.C. and Edmonds, S.J. (1972) *The Phyla Sipunculida and Echiura*. British Museum (Natural History), London. [Major taxonomic work on sipunculids and echiurans with some notes on collection and preservation.]

Index